New Frontiers in Regional Science: Asian Perspectives

Volume 34

New Frontiers in Regional Science: Asian Perspectives

This series is a constellation of works by scholars in the field of regional science and in related disciplines specifically focusing on dynamism in Asia.

Asia is the most dynamic part of the world. Japan, Korea, Taiwan, and Singapore experienced rapid and miracle economic growth in the 1970s. Malaysia, Indonesia, and Thailand followed in the 1980s. China, India, and Vietnam are now rising countries in Asia and are even leading the world economy. Due to their rapid economic development and growth, Asian countries continue to face a variety of urgent issues including regional and institutional unbalanced growth, environmental problems, poverty amidst prosperity, an ageing society, the collapse of the bubble economy, and deflation, among others.

Asian countries are diversified as they have their own cultural, historical, and geographical as well as political conditions. Due to this fact, scholars specializing in regional science as an inter- and multi-discipline have taken leading roles in providing mitigating policy proposals based on robust interdisciplinary analysis of multifaceted regional issues and subjects in Asia. This series not only will present unique research results from Asia that are unfamiliar in other parts of the world because of language barriers, but also will publish advanced research results from those regions that have focused on regional and urban issues in Asia from different perspectives.

The series aims to expand the frontiers of regional science through diffusion of intrinsically developed and advanced modern regional science methodologies in Asia and other areas of the world. Readers will be inspired to realize that regional and urban issues in the world are so vast that their established methodologies still have space for development and refinement, and to understand the importance of the interdisciplinary and multidisciplinary approach that is inherent in regional science for analyzing and resolving urgent regional and urban issues in Asia.

Topics under consideration in this series include the theory of social cost and benefit analysis and criteria of public investments, socio-economic vulnerability against disasters, food security and policy, agro-food systems in China, industrial clustering in Asia, comprehensive management of water environment and resources in a river basin, the international trade bloc and food security, migration and labor market in Asia, land policy and local property tax, Information and Communication Technology planning, consumer "shop-around" movements, and regeneration of downtowns, among others.

Researchers who are interested in publishing their books in this Series should obtain a proposal form from Yoshiro Higano (Editor in Chief, higano@jsrsai.jp) and return the completed form to him.

More information about this series at http://www.springer.com/series/13039

Keiko Nakayama • Yuzuru Miyata
Editors

Theoretical and Empirical Analysis in Environmental Economics

Springer

Editors
Keiko Nakayama
Faculty of Economics
Chukyo University
Nagoya, Aichi, Japan

Yuzuru Miyata (deceased)
Toyohashi, Aichi, Japan

ISSN 2199-5974 ISSN 2199-5982 (electronic)
New Frontiers in Regional Science: Asian Perspectives
ISBN 978-981-13-2362-1 ISBN 978-981-13-2363-8 (eBook)
https://doi.org/10.1007/978-981-13-2363-8

Library of Congress Control Number: 2019935807

This Springer imprint is published by the registered company Springer Nature Singapore Pte Ltd.
The registered company address is: 152 Beach Road, #21-01/04 Gateway East, Singapore 189721, Singapore

Preface

The main concern of people and governments of the developed countries is shifting from economic growth to life stability or environment preservation as their economies have matured.

The main purpose of this monograph is to show possible remedies on some of the current environmental issues in developed countries in theoretical and/or empirical manners with interdisciplinary approaches of economics.

This book consists of two parts, theoretical and empirical studies. The environmental issues which this book treats include forest environmental taxes. They are introduced as part of local taxes, air pollution reduction policies for mobile emission sources, introduction of renewable energies and power fuel cell technology, mechanism of city agglomeration dispersion, measurement of environment sustainability, etc. As to the analytical methods, some researches employ theoretical approaches as mathematical economic model or nonlinear dynamic model, and some analyses were implemented using empirical or statistical tools such as long-run general equilibrium model, input-output model, dynamic optimization model, and so on.

Lastly, we would like to say many thanks to the authors of papers in this book.

Nagoya, Japan Keiko Nakayama

To Celebrate the Retirement of Professor Dr. Masatoshi Shirai, Chukyo University

Professor Dr. Masatoshi Shirai is to come to an age of retirement from Chukyo University in 2019.

He enrolled in the School of Economics, Nagoya University, in 1968 and graduated in 1972. Then, he advanced to the Graduate School of Economics, Nagoya University, and studied Economics and Finance under the supervision of Professor Dr. Masaichi Mizuno. He received a doctorate degree from Nagoya University based on the achievement of theoretical studies in Education Economics.

After finishing the doctor course in 1977, he was appointed as a lecturer of Finance at the School of Commerce, Chukyo University, where his professional carrier started. He was promoted to associate professor in 1980 and moved to the School of Economics, which was newly founded at the Chukyo University in 1988. After then, he was promoted to professor of Economics in 1991 and has reached the present. Meanwhile, he contributed as the director of the Institute of Economics from 1999 to 2002 and worked as the dean of the School of Economics from 2003 to 2004 and as the dean of the Graduate School of Economics from 2005 to 2006.

His scientific concern is placed in the field of fiscal finance, public economics, and educational economics. He has taught these classes in the undergraduate and graduate School of Economics, Chukyo University, for more than 40 years. A lot of undergraduate students studied in his enthusiastic classes. He also supervised many graduate students to become academic researchers.

His research contributions were summarized as follows: *Educational economics*, Keiso Shobo, Japan (1991); *Recommendations for educational reforms by economists*, Ministry of Finance Press, Japan (1998); *Public Economics Research (III),* Keiso Shobo, Japan (2001); *Theory and challenges of asset taxation*, Taxation Accounting Association, Japan (2005); *Public Economics Studies V,* Keiso Shobo, Japan (2012); *Public Economics Studies VI*, Keiso Shobo, Japan (2017); and others. At the same time, as publishing many books in these specialized fields, many textbooks on the theory of microeconomics and fiscal theory are also published, such as *Modern Economics*, Chuo Keizaisha, Japan (2001), *The Idea of Microeconomics*, Yachiyo publication (2012), etc.

Also, as his representative academic papers, we find the following:

"The Investment Rule for Public Education" *Economics of Education Review*, 9(1), 1990

"Analysis of fiscal expenditure for higher education" *Chukyo University economic review* 13 (2002)

"The Optimal Subsidies to Higher Education with Special Reference to Education and Research" *Okayama Economic Review* 36(4), 2005

"Forest Recharge on Public Investment on Water Supply Conservation Problems – Analysis of Forest Environment Tax by Optimization Problem," *Regional Studies*, 39 (3) 2009

"Education and Research in public policy," *International Journal of Economic Behavior and Organization*, 2016, etc.

He has authored numerous papers in fields such as fiscal, public economics, and educational economics. Since he started his academic carrier at Chukyo University in 1977, he has made a great academic contribution not only to the education and research at Chukyo University but also to the Japanese economic society for longer period of over 40 years.

The authors of this book *Theoretical and Empirical Analysis in Environmental Economics* are his colleagues, friends, and researchers who have worked cooperatively with him. All of them, from their standpoint, highly appreciate his enthusiasm and academic achievement. At the same time, in commemoration of his retirement, they all are extremely pleased about their contributing together with him in this book.

I wish Professor Shirai his continued health and increasing happiness with his family after his retirement.

January 2019
As one of his colleagues and friends,
Mitsuo Yamada

Contents

Part I
Environment: Theoretical Approach

Chapter 1
The Ambient Charge in Hyperbolic Duopoly and Triopoly: Static and Dynamic Analysis

Akio Matsumoto, Keiko Nakayama, and Ferenc Szidarovszky

Abstract This paper presents a static and a dynamic model of an environmental policy of the ambient tax related to nonpoint source pollution that has many different sources. We apply the model to controllability by the ambient tax rate, showing that increasing the rate reduces the total level of the pollution. We also consider dynamic characteristics in the discrete time scales and numerically show the birth of complex dynamics via period-doubling bifurcation.

Keywords Nonpoint source pollution · Ambient charge · Hyperbolic price function · Duopoly and triopoly · Complex dynamics

1.1 Introduction

According to Xepapades (2011), nonpoint source (NPS) pollution refers to the form of pollution in which neither the source nor the size of specific emissions can be observed and identified with sufficient accuracy. The ambient concentration of pollutants associated with the individually unobserved emission in water bodies, the ground, or the air can be observed. Most types of pollutions that we can observe today are considered to be NPS. It exhibits a sharp contrast to a point source (PS) pollution that results from a single source. The regulator has enough information to control the PS pollution. It actually uses the standard instruments of environmental policy such as emission taxes, tradable emission permits, deposit-

A. Matsumoto (✉)
Department of Economics, Chuo University, Tokyo, Japan
e-mail: akiom@tamacc.chuo-u.ac.jp

K. Nakayama
Department of Economics, Chukyo University, Nagoya, Japan

F. Szidarovszky
Department of Mathematics, Corvinus University, Budapest, Hungary

K. Nakayama, Y. Miyata (eds.), *Theoretical and Empirical Analysis in Environmental Economics*, New Frontiers in Regional Science: Asian Perspectives 34, https://doi.org/10.1007/978-981-13-2363-8_1

refund system, etc. However, due to information asymmetries, the regulator cannot observe individual emissions in case of NPS pollutions. In consequence, regulating it poses serious challenges. Three approaches have been developed so far. The first approach concerns input-based instruments that are linked to observable polluting inputs. The second is called the ambient schemes that are developed to deal with the ambient concentration of the pollutants. The third approach attempts to improve information on the individual emissions. In this study, we confine our focus on the ambient scheme in the forms of an ambient tax that was first formulated by Segerson (1988). Accordingly, the regulator announces the ambient tax rate and a cutoff level of the ambient concentration and encourages the polluters to make pollution reduction by introducing associated penalties and rewards in the following way. If the deviation of the observed level from the cutoff level is positive, then the polluters pay tax proportional to the deviation. If negative, then they receive subsidy proportional to the deviation. This study considers whether the ambient tax can control NPS pollution under á la Cournot imperfect competition in which the payoff of a polluter depends on the actions of others.

In a Cournot framework in which quantity is a strategic variable, it has been demonstrated that an increase of the ambient tax can decrease the total level of NPS pollution. See Raju and Ganguli (2013) in a duopoly case and Matsumoto and Szidarovszky (2017) in an n-firm case. In the similar framework, Matsumoto et al. (2018) discuss delay dynamics in a continuous and discrete time scales. The linear price function is adopted in those papers. In this study, we replace the linear function with a nonlinear function, that is, a hyperbolic price function. Then we first look into whether such nonlinearity can affect the ambient tax effect and then discuss dynamic features of an unstable stationary point, following Puu (2003).

Section 1.2 provides a basic Cournot model with hyperbolic price and is divided into two subsections. The existence of unique Cournot equilibrium and the negative effect of the ambient tax in duopoly is discussed in the first subsection and in triopoly in the second. Section 1.3 presents dynamic analysis in discrete time scales and describes the birth of complex dynamics due to the nonlinearity. Section 1.4 contains concluding remarks and further research direction.

1.2 Static Analysis

Consider an n-firm oligopoly without product differentiation. Output of firm i is x_i, and let Q be the total demand. The price function is hyperbolic,

$$p = \frac{1}{Q} \tag{1.1}$$

where the total demand is supposed to be equal to the industry output, $\sum_{i=1}^{n} x_i$. Let c_i be marginal cost and e_i the emission technology coefficient of firm i, implying that $e_i x_i$ is the emission level. If $\bar{E}$ is the emission standard and m the penalty or reward fraction specified by the regulator,[1] then the payoff of firm i is

$$\pi_i = \frac{x_i}{\sum_{j=1}^n x_j} - c_i x_i - m\left(\sum_{j=1}^n e_j x_j - \bar{E}\right). \tag{1.2}$$

Firm i maximizes its payoff with given value of the outputs of the other firms. Solving the first-order condition

$$\frac{\sum_{j=1}^n x_j - x_i}{\left(\sum_{j=1}^n x_j\right)^2} - c_i - me_i = 0 \tag{1.3}$$

yields the best reply of firm i as

$$x_i = \sqrt{\frac{y_i}{\bar{c}_i}} - y_i \tag{1.4}$$

where y_i and $\bar{c}_i$ are the output of the rest of the industry and the extended marginal cost, which are defined as

$$y_i = \sum_{j \neq i}^n x_j \text{ and } \bar{c}_i = c_i + me_i.$$

1.2.1 Duopoly

Let us start with the simplest case in which there are two firms in the industry, $i = 1, 2$. The best reply of firm i is, from (1.4) with $n = 2$,

$$x_i = \sqrt{\frac{x_j}{c_i}} - x_j \text{ for } i, j = 1, 2 \text{ and } i \neq j. \tag{1.5}$$

Solving the best replies of firms i and j as a simultaneous system of equations for the output quantities, we obtain the Cournot equilibrium value of firm i,

$$x_i^c = \frac{\bar{c}_j}{\left(\bar{c}_i + \bar{c}_j\right)^2} = \frac{c_j + me_j}{\left[c_i + c_j + m\left(e_i + e_j\right)\right]^2} \tag{1.6}$$

[1] $\bar{E}$ and m are the strategic variables of the regulator and should be determined so as to maximize some form of a welfare function. In the present paper, these are assumed to be given exogenously; however their optimal determination will be considered in our future studies.

which assures that the Cournot output is positive. For the sake of analytical simplicity, we assume that the firms have the same production cost:

Assumption 1 $c_i = c$ for all i.

To see the effect on the Cournot value caused by a change in the policy parameter m, we first differentiate (1.6) with respect to m,

$$\frac{\partial x_i^c}{\partial m} = -\frac{2ce_i + me_j\left(e_i + e_j\right)}{\left[2c + m\left(e_i + e_j\right)\right]^3} < 0. \tag{1.7}$$

The total emission level at the Cournot equilibrium point is

$$E^C = e_i x_i^c + e_j x_j^c$$

and its derivative with respect to m is

$$\frac{\partial E^C}{\partial m} = e_i \frac{\partial x_i^c}{\partial m} + e_j \frac{\partial x_j^c}{\partial m} < 0 \tag{1.8}$$

or using (1.7) the right-hand side of (1.8) is rewritten as

$$\frac{\partial E^C}{\partial m} = -\frac{2\left[c\left(e_i^2 + e_j^2\right) + me_ie_j\left(e_i + e_j\right)\right]}{\left[2c + m\left(e_i + e_j\right)\right]^3} < 0.$$

Hence increasing the value of m has a negative effect on the total emission level. In other word, the ambient charge is effective in controlling the NPS pollution in the duopoly industry.

Theorem 1 *The ambient charge environmental policy can reduce the emission level of the NPS pollution in the Cournot duopoly framework under the hyperbolic price function and equal marginal costs.*

1.2.2 Triopoly

We increase the number of the firms to three and also consider the effect of the ambient charge. Equation (1.4) with $n = 3$ presents the best replies of the firms in the triopoly industry,

$$
\begin{aligned}
x_1 &= \sqrt{\frac{x_2 + x_3}{\bar{c}_1}} - (x_2 + x_3), \\
x_2 &= \sqrt{\frac{x_1 + x_3}{\bar{c}_2}} - (x_1 + x_3), \\
x_3 &= \sqrt{\frac{x_1 + x_2}{\bar{c}_3}} - (x_1 + x_2).
\end{aligned} \tag{1.9}
$$

Taking equations in (1.9) as a system of simultaneous equations, we obtain the optimal productions at the Cournot point,

$$
\begin{aligned}
x_1^c &= \frac{2(\bar{c}_2 + \bar{c}_3 - \bar{c}_1)}{(\bar{c}_1 + \bar{c}_2 + \bar{c}_3)^2} = \frac{2\left[c_2 + c_3 - c_1 + m(e_2 + e_3 - e_1)\right]}{\left[c_1 + c_2 + c_3 + m(e_1 + e_2 + e_3)\right]^2}, \\
x_2^c &= \frac{2(\bar{c}_1 + \bar{c}_3 - \bar{c}_2)}{(\bar{c}_1 + \bar{c}_2 + \bar{c}_3)^2} = \frac{2\left[c_1 + c_3 - c_2 + m(e_1 + e_3 - e_2)\right]}{\left[c_1 + c_2 + c_3 + m(e_1 + e_2 + e_3)\right]^2}, \\
x_3^c &= \frac{2(\bar{c}_1 + \bar{c}_2 - \bar{c}_3)}{(\bar{c}_1 + \bar{c}_2 + \bar{c}_3)^2} = \frac{2\left[c_1 + c_2 - c_3 + m(e_1 + e_2 - e_3)\right]}{\left[c_1 + c_2 + c_3 + m(e_1 + e_2 + e_3)\right]^2}.
\end{aligned}
$$

For the sake of analytical simplicity, we introduce the nonnegative variables h and k satisfying

$$
e_2 = he_1 \text{ and } e_3 = ke_1
$$

and impose Assumption 1 again. In consequence, the Cournot outputs are

$$
\begin{aligned}
x_1^c &= \frac{2me_1\,(a + h + k - 1)}{\left[3c + me_1\,(1 + k + h)\right]^2}, \\
x_2^c &= \frac{2me_1\,(a + 1 + k - h)}{\left[3c + me_1\,(1 + k + h)\right]^2}, \\
x_3^c &= \frac{2me_1\,(a + 1 + h - k)}{\left[3c + me_1\,(1 + k + h)\right]^2},
\end{aligned} \tag{1.10}
$$

where a is the ratio of the marginal production cost (i.e., c) to the marginal emission cost (i.e., me_1),

$$
a = \frac{c}{me_1} > 0.
$$

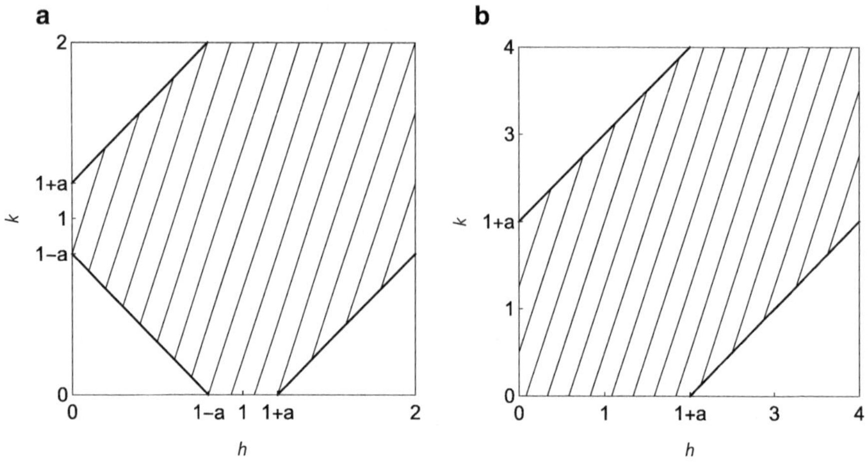

Fig. 1.1 Nonnegative output region $\boldsymbol{N}$ (**a**) $a < 1$. (**b**) $a \geq 1$

The denominator of x_i^c in (1.10) is positive. Hence, given the positive value of e_1, the feasible region of h and k in which the Cournot outputs are nonnegative is defined by

$$\boldsymbol{N} = \{(h,k) \mid h \geq 0,\ k \geq 0,\ k{+}h{-}1{+}a \geq 0,\ k{-}h{+}1{+}a \geq 0 \text{ and } h{-}k{+}1{+}a \geq 0\}.$$

Region $\boldsymbol{N}$ is described by the hatched region surrounded by one negative sloping real line (i.e., $k = -h+1-a$) and two positive sloping real lines (i.e., $k = h\pm(1+a)$) when $a < 1$ (in particular, $a = 0.2$) in Fig. 1.1a and by the two positive sloping real lines when $a \geq 1$ (in particular $a = 1$) in Fig. 1.1b.[2]

The total emission level at the Cournot point is the sum of the emission produced by each firm,

$$E^C = e_1 x_1^c + e_2 x_2^c + e_3 x_3^c > 0 \tag{1.11}$$

for $(h, k) \in \boldsymbol{N}$.[3] Substituting the optimal outputs given in (1.10) into the right hand side of (1.11) present, after arranging the terms, the following form

$$E^C = \frac{2me_1^2\,[a(1+k+h)+S_2]}{M^2} \tag{1.12}$$

where, for notational simplicity, we introduce new variables,

$$S_2 = 2(k+h+kh) - (1+k^2+h^2) \tag{1.13}$$

[2]The negative sloping line is located outside of the positive (h, k) quadrant.

[3]Notice that $x_i^c = 0$ for all i is impossible for $(h, k) \in \boldsymbol{N}$.

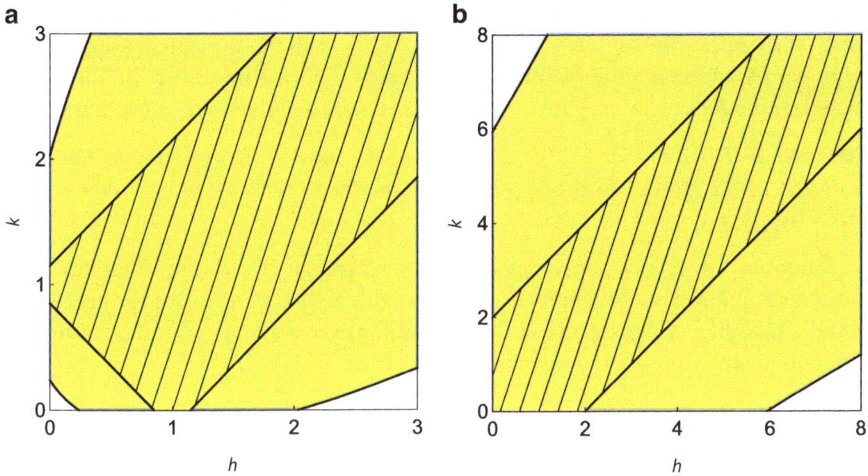

Fig. 1.2 $\partial E^C/\partial m < 0$ when $x_i^c \geq 0$ for $i = 1, 2, 3$ (**a**) $a < 1$ (**b**) $a \geq 1$

and

$$M = 3c + me_1 (1 + k + h) .$$

We now turn attention to whether changing the value of the ambient tax rate (i.e., m) can control the total level of the NPS pollution. To this end, we differentiate (1.12) with respect to m to obtain

$$\frac{\partial E^C}{\partial m} = \frac{2me_1^2}{M^3} \left[aS_1 - (1 + h + k)S_2 \right] \tag{1.14}$$

where

$$S_1 = 2(k + h + kh) - 5(1 + k^2 + h^2) \tag{1.15}$$

and notice that S_1 is negative for any values of k and h since it is rewritten as

$$S_1 = -\left[(k-1)^2 + (h-1)^2 + (k-h)^2 + 3\left(1 + k^2 + h^2\right) \right].$$

However, S_2 is not necessarily positive in region $\boldsymbol{N}$ so that the sign of $\partial E^C/\partial m$ seems to be ambiguous in general. We describe analytical consideration on the direction of the sign in the Appendix 1 but give graphical consideration in Fig. 1.2 in which the $\partial E^C/\partial m = 0$ curve (equivalently, the locus of $aS_1 - (1 + h + k)S_2 = 0$) is depicted as the three nonlinear black segments in Fig. 1.2a with $a = 0.15$ and two black upper most and lower most curves in Fig. 1.2b with $a = 1$. In both figures the region of $\partial E^C/\partial m < 0$ is illustrated in yellow where the hatched region $\boldsymbol{N}$ is superimposed. It can be seen that $\partial E^C/\partial m < 0$ in the entire hatched region. In other words, increasing the value of m decreases the total level of pollution when each firm produces a nonnegative output. This result is summarized as follows:

Theorem 2 *The ambient charge can control the total emission level of the NPS pollution in the Cournot triopoly framework under hyperbolic price function and equal marginal cost of production.*

Theorem 2 does not imply that the ambient charge is effective in controlling the emission level of the individual firms, although the regulator is unable to observe this effects. To see the individual response to a change in the tax rate, we differentiate the Cournot outputs with respect to m,

$$\begin{aligned}
\frac{\partial x_1^c}{\partial m} &= \frac{2me_1^2}{M^3}\left[a(h+k-5)-(h+k-1)(1+h+k)\right], \\
\frac{\partial x_2^c}{\partial m} &= \frac{2me_1^2}{M^3}\left[a(1+k-5h)-(1+k-h)(1+h+k)\right], \\
\frac{\partial x_3^c}{\partial m} &= \frac{2me_1^2}{M^3}\left[a(1+h-5k)-(1+h-k)(1+h+k)\right].
\end{aligned} \tag{1.16}$$

Concerning the signs of these derivatives, we have the following result, meaning that the ambient tax could have a perverse effect that an increase of the tax could increase the emission level of the individual firms.

Theorem 3 *The individual response to a change of the ambient tax rate could be positive,*

$$\begin{aligned}
&\frac{\partial x_1^c}{\partial m} > 0 \text{ if } a > a_2 \text{ and } x_1 \le k+h \le x_2, \\
&\frac{\partial x_2^c}{\partial m} > 0 \text{ if } a > a_1 \text{ and } k \ge 5h-1, \\
&\frac{\partial x_3^c}{\partial m} > 0 \text{ if } a > a_1 \text{ and } h \ge 5k-1
\end{aligned}$$

where

$$a_2 = 2(5+2\sqrt{6}),\ x_{1,2} = \frac{a \pm \sqrt{a^2-20a+4}}{2}.$$

Proof See Lemmas 1, 2 and 3 in the Appendix 2. ■

Figure 1.3a, b are illustrated with appropriate adjustment of the aspect ratio and describes regions in which $\partial x_2^c/\partial m \ge 0$ and $\partial x_3^c/\partial m \ge 0$. In particular, $\partial x_2^c/\partial m \ge 0$ in the gray regions of Fig. 1.3a in which the blue curve or the blue-red curve is

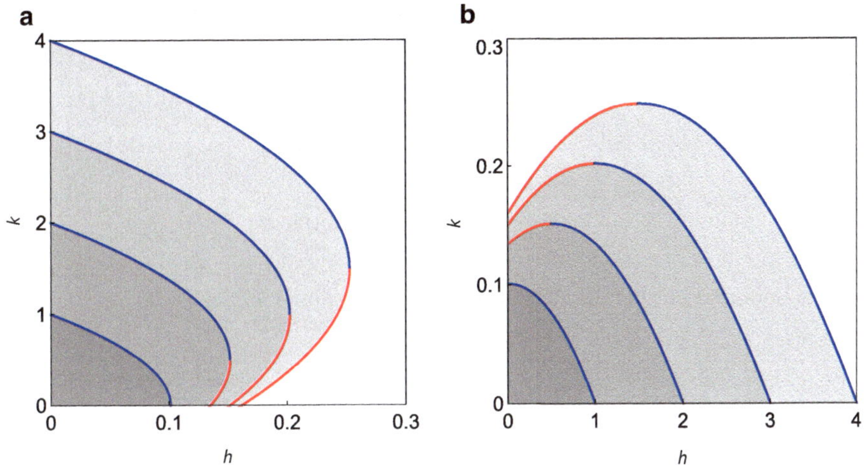

Fig. 1.3 Perverse effects of the ambient charge on individual demands (**a**) Regions for $\partial x_2^c/\partial m \geq 0$. (**b**) Regions for $\partial x_3^c/\partial m \geq 0$

the locus of $\partial x_2^c/\partial m = 0$ and the curve shifts outward as the value of a increases from $2, 3,\ 4$ and 5. As can be seen in (1.16), $\partial x_2^c/\partial m$ and $\partial x_3^c/\partial m$ are symmetric in a sense that $\partial x_i^c/\partial m$ can be obtained from $\partial x_j^c/\partial m$ by interchanging h with k for $i, j = 2, 3$ and $i \neq j$. As a natural result, the gray regions of Fig. 1.3a in which $\partial x_2^c/\partial m \geq 0$ is symmetric to the one of Fig. 1.3b in which $\partial x_3^c/\partial m \geq 0$ with respect to the diagonal of the (h, k) plane and obtained by changing the horizontal axis with the vertical axis.

Differentiating (1.11) with respect to m gives

$$\frac{\partial E^C}{\partial m} = e_1 \frac{\partial x_1^c}{\partial m} + e_2 \frac{\partial x_2^c}{\partial m} + e_3 \frac{\partial x_3^c}{\partial m}.$$

Theorem 2 states that the left-hand side of this equation is negative. Theorem 3 indicates that the derivatives of the right-hand side could be positive. However it does not say that three derivatives become positive simultaneously. Hence Theorem 2 is compatible with Theorem 3.

1.3 Dynamic Analysis

1.3.1 Duopoly Dynamics

Let us now turn to the stability problem of the Cournot point in the duopoly framework. We retain Assumption 1 in this section and, for notational simplicity, redefine the marginal cost as $c_i = c + me_i$. Assuming naive expectation formation, the best replies can be written as the iterative processes of the output adjustment,

$$\begin{aligned} x_1(t+1) &= \sqrt{\frac{x_2(t)}{c_1}} - x_2(t), \\ x_2(t+1) &= \sqrt{\frac{x_1(t)}{c_2}} - x_1(t). \end{aligned} \tag{1.17}$$

The Jacobi matrix of the discrete time dynamic system (1.17) is

$$\boldsymbol{J} = \begin{pmatrix} 0 & \dfrac{1}{2\sqrt{c_1 x_2^c}} - 1 \\ \dfrac{1}{2\sqrt{c_2 x_1^c}} - 1 & 0 \end{pmatrix} = \begin{pmatrix} 0 & -\dfrac{c_1 - c_2}{2c_1} \\ \dfrac{c_1 - c_2}{2c_2} & 0 \end{pmatrix}$$

where the elements are evaluated at the Cournot point. The characteristic equation is

$$|\boldsymbol{J} - \lambda \boldsymbol{I}| = \lambda^2 + \frac{(c_1 - c_2)^2}{4c_1c_2} = 0.$$

The stability condition that the roots of the characteristic equation are less than unity in absolute value is spelled out as

$$\frac{(c_1 - c_2)^2}{4c_2c_2} < 1$$

or

$$3 - 2\sqrt{2} < \frac{c_1}{c_2} < 3 + 2\sqrt{2}.$$

This conditions imply that Cournot equilibrium is locally asymptotically stable if the firms are *similar* to the extent that the ratio of the marginal costs stays in the designated interval. Returning to the definitions of c_1 and c_2, we obtain the following conditions for stability:

Theorem 4 *The Cournot point is locally asymptotically stable if the emission parameters satisfy*

$$2\left(1 - \sqrt{2}\right)\frac{c}{m} + (3 - 2\sqrt{2})e_1 < e_2 < 2\left(1 + \sqrt{2}\right)\frac{c}{m} + (3 + 2\sqrt{2})e_1 \tag{1.18}$$

The yellow region in Fig. 1.4a corresponds to the stability region in which the upper boundary is described by

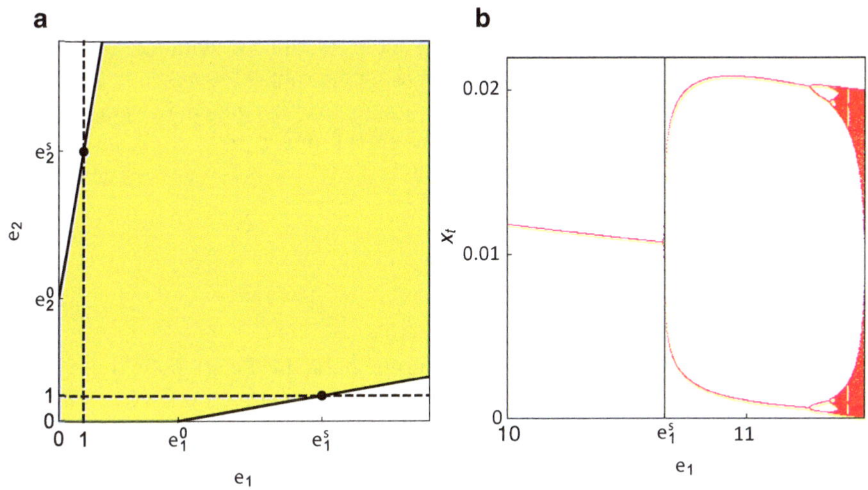

Fig. 1.4 Stability region and time evolution of x_t (**a**) Stability region. (**b**) Bifurcation diagram

$$e_2 = e_2^0 + (3 + 2\sqrt{2})e_1$$

and the lower boundary by

$$e_1 = e_1^0 + (3 + 2\sqrt{2})e_2$$

where the horizontal and vertical intersections take the same value,

$$e_1^0 = e_2^0 = 2\left(1 + \sqrt{2}\right)\frac{c}{m}.$$

This is, needless to say, equivalent to

$$e_j = 2\left(1 + \sqrt{2}\right)\frac{c}{m} + (3 + 2\sqrt{2})e_i,\ i \neq j.$$

Although the parameters are specified as $c = m = 1$ in Fig. 1.4a, the shape of the stability region is essentially the same for any values of c and m. The Cournot point can be locally unstable if firms strongly asymmetric in the sense that inequality condition (1.18) is violated, that is, a pair of (e_1, e_2) is in the white region. Further, it is now well-known that the dynamic system (1.17) can give rise to complex dynamic involving chaos through a period doubling cascade when the stability of the Cournot point is violated (e.g., see Puu 2003). Indeed, the bifurcation diagram of x_1^c is illustrated in Fig. 1.4b in which the initial point is taken as $e_2^0 = 1$ and the value of e_1 increases along the horizontal (vertical) dotted line of Fig. 1.4a. The stability of the stationary point is lost at $e_1^s = 5 + 4\sqrt{2}$, and the stationary point bifurcates to a period-2 cycle, which finally exhibits chaotic oscillations after going through a period-doubling bifurcation. Exactly the same bifurcation diagram of x_2^c with respect to e_2 is obtained by replacing the subscript "1" with "2." Instability means that the environmental policy of increasing the value of m may not lead to the expected result of a decrease of the total emission level but generates various types of ups and downs of the production levels of each firm according to the strength of the emission parameters.

1.3.2 Triopoly Dynamics

We next consider the stability of the Cournot point in the triopoly framework. As in the duopoly dynamics, we assume the naive expectation formation and differentiating the best reply functions to obtain the Jacobian matrix,

$$\boldsymbol{J} = \begin{pmatrix} 0 & \dfrac{1}{2\sqrt{c_1(x_2^c + x_3^c)}} - 1 & \dfrac{1}{2\sqrt{c_1(x_2^c + x_3^c)}} - 1 \\ \dfrac{1}{2\sqrt{c_2(x_1^c + x_3^c)}} - 1 & 0 & \dfrac{1}{2\sqrt{c_2(x_1^c + x_3^c)}} - 1 \\ \dfrac{1}{2\sqrt{c_3(x_1^c + x_2^c)}} - 1 & \dfrac{1}{2\sqrt{c_3(x_1^c + x_2^c)}} - 1 & 0 \end{pmatrix}.$$

Since the J-matrix is evaluated at the Cournot point, we substitute the coordinates for the Cournot point to get the simplified matrix,

$$\boldsymbol{J} = \begin{pmatrix} 0 & \dfrac{c_2 + c_3 - 3c_1}{4c_1} & \dfrac{c_2 + c_3 - 3c_1}{4c_1} \\ \dfrac{c_1 + c_3 - 3c_2}{4c_2} & 0 & \dfrac{c_1 + c_3 - 3c_2}{4c_2} \\ \dfrac{c_1 + c_2 - 3c_3}{4c_3} & \dfrac{c_1 + c_2 - 3c_3}{4c_3} & 0 \end{pmatrix}.$$

The characteristic equation becomes cubic,

$$|\boldsymbol{J} - \lambda \boldsymbol{I}| = -\lambda^3 + A\lambda + B = 0$$

where

$$A = \frac{6(c_1^3 + c_2^3 + c_3^3) - 5(c_1^2 + c_2^2 + c_3^2)(c_1 + c_2 + c_3) + 30c_1c_2c_3}{16c_1c_2c_3}$$

and

$$B = \frac{(c_1 + c_2 - 3c_3)(c_1 + c_3 - 3c_2)(c_2 + c_3 - 3c_1)}{32c_1c_2c_3}.$$

In Farebrother (1973), stability condition for the cubic equation,

$$x^3 + a_1 x^2 + a_2 x + a_3 = 0,$$

are

$$1 + a_2 > |a_1 + a_3|,$$

$$1 - a_2 + a_1 a_3 - a_3^2 > 0$$

$$a_2 < 3.$$

Our equation is $\lambda^3 - A\lambda - B = 0$, so $a_1 = 0,\ a_2 = -A$ and $a_3 = -B$. Hence the stability conditions becomes

$$1 - A > |B|,$$

$$1 + A - B^2 > 0$$

and

$$A > -3.$$

We are now ready to express the condition for gaining stability in our own notation by replacing c_j with $c + me_j$ for $j = x, y, z$ and then using h and k

$$A = \frac{30(a+1)(a+h)(a+k) - 5[3a+(1+h+k)]\left[(a+1)^2+(a+h)^2+(a+k)^2\right] + 6\left[(a+1)^3+(a+h)^3+(a+k)^3\right]}{16(a+1)(a+h)(a+k)}$$

and

$$B = \frac{[a-(1+h-3k)]\,[a+(3-h-k)]\,[a-(1-3h+k)]}{32(a+1)(a+h)(a+k)},$$

where notice that $a = c$ under the specified parameter values. Accordingly, the positive quadrant of the (h, k) plane is divided into the stability region and the instability region by the curves of $B^2 - A = 1$, $A + B = 1$ and $A - B = 1$.[4] Since it is too complicate to consider the stability condition in an analytic way, we graphically examine it by illustrating the region divisions with the specified values of the parameters, $e_1 = 1,\ m = 1$ and $c = 0.1$ in Fig. 1.5a and $c = 1$ in Fig. 1.5b. In both figures, $B^2 - A < 1$ is satisfied in the yellow region, $A + B < 1$ is satisfied in the yellow and green regions, and $A - B < 1$ is satisfied in the yellow, green, and orange regions. Hence all three conditions are satisfied in the yellow region, which

[4]Condition $A = -3$ does not affect the stability region and is omitted.

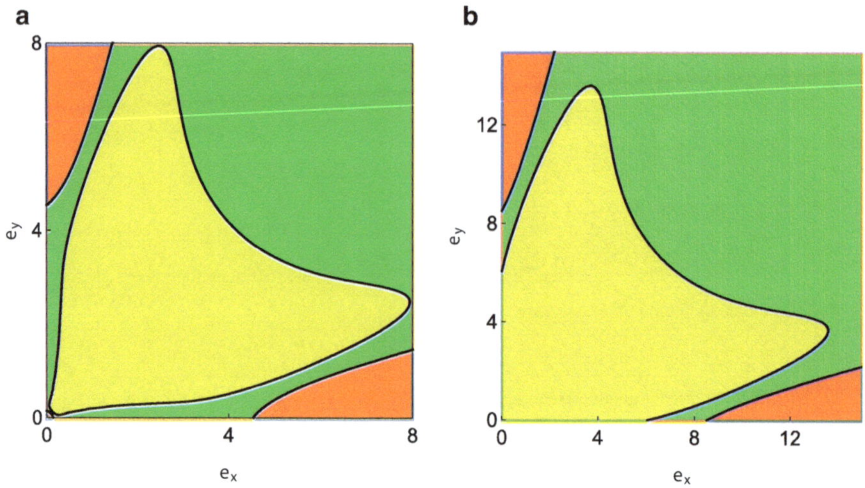

Fig. 1.5 The stability and instability regions in the triopoly case (**a**) $c = 0.1$. (**b**) $c = 1$

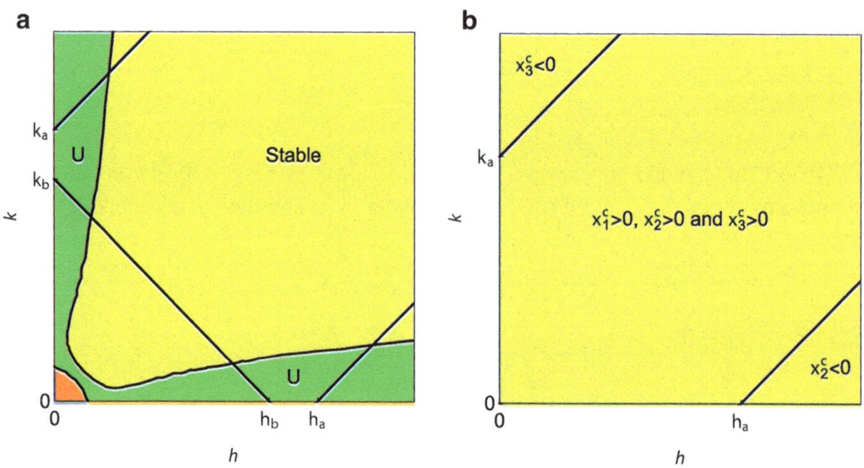

Fig. 1.6 Stable region with positive Cournot outputs (**a**) $a = 0.1$. (**b**) $a = 1$

is thus the stability region. Comparing the yellow regions in both figures implies that increasing the value of c enlarges the yellow region not only in the first quadrant but also in the second, third, and fourth quadrants, which are, however, eliminated by the non-negativity conditions of e_1 and e_1 in Fig. 1.5b. Needless to say, the green and orange regions are also changed accordingly.

We are now ready to consider the stability conditions and the non-negativity conditions together. Figure 1.6a is an enlargement of the lower left part of Fig. 1.5a in which the distorted L-shaped curve is described by $B^2 - A = 1$ and the three lines correspond to the loci of $x_1^c = 0$, $x_2^c = 0$ and $x_3^c = 0$. The intercepts are $h_a = k_a = 1 + a$ and $h_b = k_b = 1 - a$. In the yellow region (indicated with "stable"), the Cournot outputs are positive and locally asymptotically stable, and thus increasing m decreases the total emission level and brings the economy to the new Cournot point. On the other hand, in the green regions (indicated with "U"), the Cournot outputs are positive but unstable, implying that increasing m disturbs the economy and does not guarantee the restoration of the Cournot point. Figure 1.6b is an enlargement of the lower left part of Fig. 1.5b in which the Cournot point is stable in the whole region. The upper black line is the $x_3^c = 0$ line and $x_3^c > 0$ below the line. The lower black line is the $x_2^c = 0$ line and $x_2^c > 0$ above the line. With $c = 1$, the $x_1^c = 0$ line is located outside the yellow region, and thus $x_1^c > 0$ in this region. The regions with negative productions are eliminated for further considerations. Hence in the region between the upper and lower lines, the Cournot outputs are positive and stable. Therefore the environmental policy is effective in this region.

1.4 Concluding Remarks

This paper presents a simple model of an environmental policy with two specific purposes. One concerns with static analysis to provide a plausible account of controllability of the ambient charge on NPS pollution. The other concerns with dynamic analysis to see how the nonlinearity of the price function affects dynamic characteristics, especially when stability is lost. It is demonstrated that the ambient charge can regulate NPS pollution for which the standard environmental policy instruments are inappropriate to apply. It is also demonstrated that complex dynamics arises through period-doubling bifurcation because the nonlinearity of the model prevents unstable trajectories from moving globally away from the stationary point.

One direct extension is to increase the number of the firms to four and more. Further, in the current analysis, the abatement technology, the ambient tax rate, and the cutoff level are exogenously given. However, these are choice variables of the firms and the regulator. In future studies, we will conduct research on how the firms and the regulator can select these variables optimally.

Appendix 1

In this Appendix, we determine the sigh of the derivative of E^c with respect to m. Equation (1.14) implies that the sign of the derivative is the same as the sign of $aS_1 - (1 + k + h)S_2$. It is repeated below as (1.19) for convenience,

$$sigh\left[\frac{\partial E^c}{\partial m}\right] = sigh\left[aS_1 - (1 + k + h)S_2\right]. \tag{1.19}$$

In this appendix, we derive the conditions under which $aS_1 - (1+k+h)S_2 < 0$. It is already shown that S_1 is negative. Hence if $S_2 \geq 0$, then the sign is definitely negative. So we assume that $S_2 < 0$.

We will now prove that the ambient charge is effective in controlling the NPS pollution. We can always number the firms so that

$$e_1 \geq e_2 \geq e_3$$

or

$$1 \geq h \geq k.$$

From the definition of the feasible set $\boldsymbol{N}$, the nonnegative conditions of the Cournot outputs can be rewritten as

$$a \geq 1 - k - h, \ a \geq h - k - 1 \text{ and } a \geq k - h - 1$$

from which the first condition is the only meaningful if $k + h < 1$ which is assumed first.[5] Then we have

$$aS_1 - (1 + k + h)S_2 = S_1(a - S) \tag{1.20}$$

where

$$S = \frac{(1+k+h)S_2}{S_1} > 0 \text{ as } S_1 < 0 \text{ and } S_2 < 0.$$

If $S < 1 - k - h$, then the second factor of the right-hand side of (1.20) is positive,

$$a - S > a - (1 - k - h) \geq 0$$

where the last inequality is due to the non-negativity condition of $x_1^c \geq 0$. Since $S_1 < 0$, (1.19) implies $\partial E^c/\partial m < 0$. Thus our job is to show that $S < 1 - k - h$ always.

[5]Suppose that $h \geq 1 \geq k$. Letting $k = \bar{k}h$, $h = \bar{1}h$, and $1 = \bar{h}h$ leads to the new ordering $\bar{1}h \geq \bar{h}h \geq \bar{k}h$. Dividing it by h and then dropping the bars from $\bar{1}$, $\bar{h}\,\bar{k}$ yield $1 \geq h \geq k$. In the same way, the non-negativity condition of x_2^c, $a \geq h - k - 1$ can be written as

$$\bar{a} \geq \bar{1} - \bar{k} - \bar{h}$$

if $a = \bar{a}h$. Notice that the non-negativity condition of x_2^c in new parameters is exactly the same as the non-negativity condition of x_1^c in the original parameters. Dropping the bars has no problem. Any other conditions can be transformed in the same way. Thus considering the case $1 \geq h \geq k$ is enough.

Since $S_1 < 0$, then this inequality can be rewritten as

$$(1 + k + h)S_2 > (1 - k - h)\, S_1$$

or

$$-2\left[3\left(k^3 + h^3\right) - 4\left(k^2 + h^2\right) + 3\,(k + h) + kh\,(k - 4 + h) - 2\right] > 0. \quad (1.21)$$

Since the condition $S_2 < 0$ can be rewritten as

$$2\left[2\,(k + h + kh) - (h^2 + k^2)\right] < 2,$$

the square bracketed terms of (1.21) is larger than

$$3\left(k^3 + h^3\right) - 4\left(k^2 + h^2\right) + 3\,(k + h) + kh\,(k - 4 + h) - 2\left[2\,(k + h + kh) - (h^2 + k^2)\right]$$

which is, after arranging the terms,

$$3\left(k^3 + h^3\right) - \left(k^2 + h^2\right) - (k + h) + k^2\,(h - 1) + h^2\,(k - 1) - 8kh.$$

Equation (1.21) is now written as

$$2\left[k^2\,(1 - h) + h^2\,(1 - k) - 3\left(k^3 + h^3\right) + \left(k^2 + h^2\right) + (k + h) + 8kh\right]. \quad (1.22)$$

However $1 - k > h$ and $1 - h > k$, (1.22) is larger than

$$2\left[-2\left(k^3 + h^3\right) + \left(k^2 + h^2\right) + (k + h) + 8kh\right]$$

$$= 2\left[k^2(1 - k) + k\left(1 - k^2\right) + h^2(1 - h) + h\left(1 - h^2\right) + 8kh\right] > 0.$$

Hence $S < 1 - k - h$ holds if $k + h < 1$. If $k + h \geq 1$, then

$$S_2 = 2(h + k)^2 - (h - k)^2 - 1 \geq 0$$

since $h \leq 1$ and $k \leq 1$ are assumed. Hence $\partial E^c / \partial m$ is always negative.

Appendix 2

In this Appendix, we examine the signs of the derivatives in (1.16) and show that the ambient charge could have perverse effect on the individual level of emission.

Lemma 1 $\dfrac{\partial x_1^c}{\partial m} \geq 0$ *is possible if* $a \geq a_2$ *and* $x_1 \leq k+h \leq x_2$ *where*

$$a_2 = 2(5+2\sqrt{6}),$$

$$x_1 = \frac{a - \sqrt{a^2 - 20a + 4}}{2}$$

and

$$x_2 = \frac{a + \sqrt{a^2 - 20a + 4}}{2}.$$

Proof A new variable $x = k + h$ is introduced. Then the square bracketed terms of the first equation of (1.16) can be rewritten as

$$f(x) = -x^2 + ax + (1 - 5a)$$

implying that

$$sigh\left[\frac{\partial x_1^c}{\partial m}\right] = sigh\left[f(x)\right].$$

Solving $f'(x) = 0$ presents $x_m = a/2$ and the corresponding maximum value, $f(x_m) = D/4$ with $D = a^2 - 20a + 4$. It is checked that $D = 0$ is attained at the following two points,

$$a_1 = 2\left(5 - 2\sqrt{6}\right) \simeq 0.202 \text{ and } a_2 = 2\left(5 + 2\sqrt{6}\right) \simeq 19.798$$

indicating that $D \geq 0$ for $a \leq a_1$ or $a \geq a_2$ and $D < 0$ for $a_1 < a < a_2$. Solving $f(x) = 0$ yields two solutions,

$$x_1 = \frac{a - \sqrt{D}}{2} \text{ and } x_2 = \frac{a + \sqrt{D}}{2}.$$

Taking account of $f(0) \gtreqless 0$ depending on $a \lesseqgtr 1/5$, we then identify four cases according to the value of a, $0 < a < 1/5$, $1/5 \leq a \leq a_1$, $a_1 < a < a_2$ and $a \geq a_2$.

(i) In case of $0 < a < 1/5$, we have $x_1 < 0$ and $x_2 > 0$. Hence, $f(x) \geq 0$ for $0 \leq x \leq x_2$ and $f(x) < 0$ for $x > x_2$. Returning to the definition of x, $x \leq x_2$ can be rewritten as

$$k + h \leq \frac{a + \sqrt{D}}{2}.$$

On the other hand, the non-negativity condition of x_1^c is given by

$$k + h \geq 1 - a.$$

Since it can be confirmed that

$$(1 - a) - \frac{a + \sqrt{D}}{2} > 0 \text{ for } a < 0.2,$$

the condition $x \leq x_2$ does not satisfy the non-negativity so that we can eliminate it for further consideration. Hence we have $f(x) < 0$ and therefore $\partial x_1^c / \partial m < 0$ for $x > x_2$.

(ii) In case of $1/5 \leq a \leq a_1$, we have $0 \leq x_1 < x_2$, leading to the following,

$$\begin{aligned} &(\mathrm{ii}-\mathrm{a})\ f(x) \leq 0 \text{ for } 0 \leq x \leq x_1, \\ &(\mathrm{ii}-\mathrm{b})\ f(x) \geq 0 \text{ for } x_1 \leq x \leq x_2, \\ &(\mathrm{ii}-\mathrm{c})\ f(x) < 0 \text{ for } x > x_2. \end{aligned}$$

Under the condition of $0.2 \leq a \leq a_1$, we can confirm that

$$1 - a > x_2 > x_1.$$

Hence x satisfying $0 \leq x \leq x_2$ violates the non-negativity condition. Therefore we can have only the last case.

(iii) In case of $a_1 < a \leq a_2$, $D \leq 0$ where the equality holds only when $a = a_2$. Hence $f(x) < 0$, so $\partial x_1^c / \partial m < 0$ for $x_1 < x < x_2$.

(iv) In case of $a > a_2$, we have $0 < x_1 < x_2$ and then

$$\begin{aligned} &(\mathrm{iv}-\mathrm{a})\ f(x) \leq 0 \text{ for } 0 \leq x \leq x_1, \\ &(\mathrm{iv}-\mathrm{b})\ f(x) \geq 0 \text{ for } x_1 \leq x \leq x_2, \\ &(\mathrm{iv}-\mathrm{c})\ f(x) < 0 \text{ for } x > x_2. \end{aligned}$$

Summarizing the results, we have

$$\frac{\partial x_1^c}{\partial m} \geq 0 \text{ is possible if } a \geq a_2 \text{ and } x_1 \leq k + h \leq x_2.$$

This completes the proof. ■

We now proceed to the second equation of (1.16).

Lemma 2 $\dfrac{\partial x_2^c}{\partial m} \geq 0$ *is possible if* $a \geq 1$ *and* $k \geq 5h - 1$.

Proof We denote by $g_1(h, k)$ the square bracketed terms of the second equation of (1.16),

$$g_1(h, k) = a(1 + k - 5h) - (1 + k - h)(1 + h + k)$$

where

$$sigh\left[\frac{\partial x_2^c}{\partial m}\right] = sigh\left[g_1(h, k)\right].$$

The non-negativity condition of x_2^c is $a \geq h - k - 1$. We identify the following three cases, (i) $h - k - 1 \geq 0$, (ii) $h - k - 1 < 0$ with $1 + k - 5h \leq 0$, and (iii) $1 + k - 5h > 0$ under which $h - k - 1 < 0$ always holds.

(i) Since $-(1+k-h) \geq 0$ is assumed, multiplying both sides of the non-negativity condition by $1 + h + k$ presents

$$a\,(1 + k + h) \geq -(1 + k - h)(1 + h + k) \geq 0.$$

The first inequality implies that the bracketed term is less than or equal to

$$a(1 + k - 5h) - a\,(1 + k + h) = 2a(1 + k - 2h) < 0$$

where the last inequality is due to the assumption $h - k - 1 \geq 0$ or $1 + k - h \leq 0$. Therefore $g_1(h, k) < 0$ for any $a \geq 0$ if $h - k - 1 \geq 0$.

(ii) Now suppose that $h - k - 1 < 0$ with which the non-negativity condition is always satisfied. The second term of the bracketed terms is negative. So if $1 + k - 5h \leq 0$, then $g_1(h, k) < 0$ for any $a \geq 0$.

(iii) Assume finally that $1 + k - 5h > 0$. Then $g_1(h, k)$ is a linear function of a with a positive slope. The non-negativity condition for the equilibrium outputs are satisfied if

$$a \geq \max\{h - 1 - k;\ 1 - h - k;\ k - 1 - h\} = a^*.$$

The value of $g_1(h, k)$ is zero if

$$a = \Omega(h, k) = \frac{(1 + k - h)(1 + h + k)}{1 + k - 5h}.$$

We can next prove that $\Omega(h, k) > 1$, which can be written as

$$(1 + k - h)(1 + h + k) - (1 + k - 5h) > 0.$$

The simplified left-hand side satisfies

$$
\begin{aligned}
&1+k+h+k+kh+k^2-h-h^2-kh-1-k+5h\\
&=k+k^2+5h-h^2\\
&=k(k+1)+5h-h^2\\
&>(5h-1)\,5h+5h-h^2\\
&=24h^2>0.
\end{aligned}
$$

Therefore $g_1(h,k)>0$ with positive equilibrium output value if

$$a>\max\left\{a^*,\ \Omega(h,k)\right\}$$

which completes the proof. ■

Notice that the sign of $a-\ \Omega(h,k)>1$ is the same as that of

$$
\begin{aligned}
&a+ak-5ah-\left(1+k-h+k+kh+k^2-h-h^2-kh\right)\\
&=-k^2+k(a-2)+\left(a-5ah-1+h^2\right)
\end{aligned}
$$

which is a concave parabola with roots

$$k_1=f_1(h)=\frac{1}{2}\left[-2+a-\sqrt{a^2-20ah+4h^2}\right]$$

and

$$k_2=f_2(h)=\frac{1}{2}\left[-2+a+\sqrt{a^2-20ah+4h^2}\right]$$

In Fig. 1.3a, four blue curves (i.e., $k_2=f_2(h)$) and three red curves (i.e., $k_1=f_1(h)$) are illustrated for $a=2,3,4,5$ where the curves shift outward as a increases. Notice that $a>\Omega(h,k)$ holds in each gray region, that is, $g_1(h,k)>0$.

The derivative $\partial x_3^c/\partial m$ can be obtained from $\partial x_2^c/\partial m$ by interchanging h and k, so from Lemma 2, we have the following result.

Lemma 3 $\dfrac{\partial x_3^c}{\partial m}\geq 0$ *is possible if* $a>1$ *and* $h>5k-1$.

As in the previous case, we can have the roots of equation $a-\Omega(k,h)$ as

$$h_1=f_1(k)=\frac{1}{2}\left[-2+a-\sqrt{a^2-20ak+4k^2}\right]$$

and

$$h_2=f_2(k)=\frac{1}{2}\left[-2+a+\sqrt{a^2-20ak+4k^2}\right]$$

where are illustrated in Fig. 1.3b.

These lemmas lead to Theorem 3.

References

Farebrother RW (1973) Simplified Samuelson conditions for cubic and quartic equations. Manchester Sch Econ Soc Stud 41:396–400
Matsumoto A, Szidarovszky F (2017) Environmental effects of ambient charge in Cournot oligopoly. J Environ Econ Policy. https://doi.org/10.1080/21606844.2017.1347527
Matsumoto A, Szidarovszky F, Takizawa H (2018) Extended oligopolies with pollution penalties and rewards. Discret Dyn Nat Soc. https://doi.org/10.1155/2018/786432
Puu T (2003) Attractors, bifurcations, and chaos, 2nd edn. Springer, Berlin/Heidelberg
Raju S, Ganguli S (2013) Strategic firm interaction, returns to scale, environmental regulation and ambient charges in a Cournot duopoly. Technol Invest 4:113–122
Segerson K (1988) Uncertainty and incentives for non-point pollution control. J Environ Econ Manag 15:87–98
Xepapades A (2011) The economics of nonpoint source pollution, Ann Rev Resour Econ 3:335–374

Chapter 2
Free-Entry Cournot Oligopoly, Environmental Policies, and the Role of Public Enterprises in a Mixed Economy

Xin Zhi Yu and Yasuyuki Nishigaki

Abstract Previous studies have noticed that environmental policy in a free-entry oligopolistic market crucially depends on the excess supply due to existence of external diseconomy, production decrease according to imperfect competition, and excessive entry into the industry. This paper examines the environmental policies and the sustainability of the industry in a free-entry Cournot-Nash oligopoly mixed market with welfare-maximizing public firms. Furthermore, we will investigate the effects of an emission abatement technology induction policy of the public firm. Our results show that the equilibrium level of production and the number of firms are smaller in the case of total emission tax than the proportional emission tax which suggests that the total emission tax alleviates the possible inefficiency of the free-entry oligopolistic market. Second, public production alleviates the inefficiency by increasing the output and decreasing the possible excess entry of private firms. Lastly, introduction of cleaner technology by the public firm may improve efficiency by inducing less production by firms with inferior production technology.

Keywords Free entry · Environmental tax · Abatement technology · Mixed economy

JEL Classification L12; H23; Q58

2.1 Introduction

Environmental protection policy has been recognized as one of the most important government actions to maintain socioeconomic sustainability. Investigations have been made in the field of environmental economics based on market mechanisms

X. Z. Yu · Y. Nishigaki (✉)
Graduate School of Economics and Faculty of Economics, Ryukoku University, Kyoto, Japan
e-mail: nisigaki@econ.ryukoku.ac.jp

K. Nakayama, Y. Miyata (eds.), *Theoretical and Empirical Analysis in Environmental Economics*, New Frontiers in Regional Science: Asian Perspectives 34, https://doi.org/10.1007/978-981-13-2363-8_2

and the representative measure of environmental tax or the emission permit trade, which constitutes incentive-compatible policy.

The pioneering work on environmental tax by Pigou (1920) argued that a tax imposed according to the cost of environmental discharge achieves the optimal resource allocation in the market. However, the so-called Pigouvian tax is known to be effective only under perfect competition, and it suggests that strategic behavior of producers leads to different consequences under imperfect market competition.

Davis and Whinston (1962), Barnet (1980), Damania (1996), and Fujiwara (2009) investigated market equilibrium under monopolistic or oligopolistic competition with external diseconomy associated with production. In an imperfect market, the level of production and total supply are generally less than those in a perfect market, and the efficient level of production depends on both production decrease according to imperfect competition and excess supply due to the existence of external diseconomy.

In the free-entry market, Mankiw and Whinston (1986) demonstrated that so-called business stealing effects under imperfect competition induced less production, but excessive entry causes welfare loss in the market. In the field of environmental economics, Katsoulacos and Xepapadeas (1995) and Lee (1999) showed that in a free-entry imperfect competition market, environmental tax causes a decrease in production and excessive entry, which leads to a decrease in efficiency. The optimal environmental tax should be set at the level of marginal environmental damage because the production decrease and excessive entry effects of the tax are offset.

In the literature on mixed markets, which is comprised of both private and public firms, Matsumura and Kanda (2005), Brandao and Castro (2007), Capuano and De Feo (2010), Ino and Matsumura (2010), Cato and Matsumura (2015), and Lili and Sang-Ho (2018) examined the possible advantages of public ownership in the free-entry market. They found that a welfare-maximizing public firm can increase social welfare by reducing the number of private firms entering the market and prevent excessive entry effects. However, inefficiency caused by less production remains the same.

This paper examines the environmental policies and the sustainability of the industry in a free-entry Cournot-Nash oligopoly mixed market with welfare-maximizing public firms. Furthermore, we will investigate the effects of an emission abatement technology induction policy implemented by the public firm.

The contribution of this paper is twofold. First, we will introduce a total emission charge environmental tax and study the ease of production decrease caused by the oligopolistic market competition. Second, we will focus on the introduction of a high-cost but clean production technology by the public firm and study the efficiency effects of market equilibrium.[1]

[1]In this paper, we construct a simple model based on Lili and Sang-Ho (2018). However, we will introduce an environmental tax on total emission and higher-cost and cleaner production technology into the basic model.

The main findings are summarized as follows. The equilibrium level of each firm's production and the number of firms are smaller in the case of total emission tax than the proportional emission tax levied on the level of production, which suggests that the introduction of the total emission tax alleviates the possible inefficiency of the free-entry oligopolistic market. Second, introduction of public production has a certain role in alleviating the inefficiency by increasing the output and decreasing the possible excess entry of private firms. Lastly, introduction of higher-cost and cleaner technology by the public firm may improve efficiency by inducing less production by firms with inferior production technology.

This paper is structured as follows. In Sect. 2.2, the basic model of a free-entry oligopolistic market with production externality is developed, and the basic solutions are presented. In Sect. 2.3, the total emission tax and its effects on the market equilibrium are studied. In Sect. 2.4, we extend the model into a mixed economy with public enterprise and study the effects of public production. Finally, in Sect. 2.5, we introduce higher-cost and cleaner production technology of the public firm and investigate the possible welfare improvement. Section 2.6 concludes the paper.

2.2 The Basic Model

This section presents the main features of the basic model. We consider a simplified industry that confronts a Cournot-Nash oligopoly market consisting of many firms. Entry into or withdrawal from the market is free, but it incurs a certain amount of fixed cost.

2.2.1 *Production Process*

We assume a simplified production process. There are a total of n firms in the market. All firms are assumed to have the same production function and produce the same products. The inversed demand function is shown as

$$P = 1 - Q. \tag{2.1}$$

where Q is the total supply to the market and is shown as

$$Q = \sum_{j=1}^{n} q^{j} \tag{2.2}$$

where the suffix $j = 1, 2, \ldots, n$ is applied to the number of the firms and q^j is the output of the jth firm.

As previously stated, all the firms have the same cost function, which is shown as

$$c\left(q^j\right) = \frac{\left(q^j\right)^2}{2} + F^2. \tag{2.3}$$

where F^2 is entry cost and is applied uniformly to all the firms.

In the industry, external diseconomy is emissions associated with the production. The level of emission μ^j is proportional to the output level of each firm q^j. The total environmental damage is shown as

$$ED = \frac{\left(\sum_{j=1}^{n} \mu^j\right)^2}{2}. \tag{2.4}$$

The government imposes an emission tax on the emission level with the tax rate t.

Given the setting of the market structure, consumer's surplus is shown as

$$CS = \frac{Q^2}{2} - ED. \tag{2.5}$$

As the level of emission damage is proportional to the level of output, the consumers' surplus increases with the output level.

2.2.2 Equilibrium of the Market

We consider a three-stage Cournot-type oligopoly game. In the first stage, the government sets the emission tax rate $T = t\mu^i = tq^i$ before the firm's entry. In the second stage, firms decide whether they enter the market and produce the goods according to the expected profit given the emission tax rate. Finally, in the third stage, each firm decides their level of production in the Cournot-Nash equilibrium.

The profit function of the firm is shown as

$$\Pi^i = Pq^i - \frac{\left(q^i\right)^2}{2} - t\mu^i - F^2, \quad (i = 1, 2, \ldots, n) \tag{2.6}$$

The sub-game perfect Nash equilibrium is solved through the recursive induction method as follows.

In the third stage, the optimal level of production is obtained according to the profit maximization of each firm. Due to the assumptions of identical production technology and the cost of entry, possible Cournot-Nash equilibrium is symmetrical.

By substituting Eqs. (2.1) and (2.2) into Eq. (2.6) and solving the profit maximization problem, the optimal level of production is obtained in the function of the emission tax rate and the number of the firms as follows.

$$q^i = \frac{1-t}{n+2} \tag{2.7}$$

Profit of each firm is shown by substituting Eq. (2.7) into (2.6) as follows.

$$\Pi^i = \frac{(1-t)^2(1+2n)}{2(n+2)^2} - F^2 \tag{2.8}$$

In the second stage, we calculate the equilibrium of the industry. Due to the assumption of free entry and withdrawal, the profit of each firm is zero in the Cournot-Nash equilibrium. By setting Eq. (2.8) as equal to zero, we obtain the equilibrium number of the firms under the equilibrium as follows.

$$n = \frac{(1-t)\left(1-t-F\sqrt{2}\right)+2F\sqrt{t}}{2F^2} - 2 \tag{2.9}$$

Equation (2.9)implies that the equilibrium number of the firms depends on the entry cost and the emission tax rate and that the equilibrium number of the firms decreases (increases) according to the increase (decrease) in the cost of entry and that of the emission tax rate.

Finally, in the first stage, the government sets the emission tax rate according to the policy objectives.[2]

2.3 Case of Total Emission Tax

In this section, we examine effects of total emission tax, which is an alternative measure for controlling the emissions of the externality and is imposed according to the total emission of the external diseconomy in the industry. Here we assume that the government sets the total emission tax rate $T = \sum_{i=1}^{n}\left(s\sum_{i=0}^{n} qi\right)$ in advance of the entries of private firms. Given the tax rate, private and public firm compete against each other in Cournot-Nash-type oligopolistic competition behavior.

Here, the profit function of private firms is shown as

$$\Pi^i = Pq^i - \frac{\left(q^i\right)^2}{2} - s\sum\nolimits_{i=0}^{n} qi - F^2 \quad (i = 0, 1, \ldots, n) \tag{2.10}$$

In this section, the sub-game perfect Nash equilibrium can be solved by the recursive induction method as follows.

[2]Here, we consider the simple method of setting the tax rate. In Sect. 2.4, we assume that the government sets the optimal tax rate to maximize social welfare.

In the third stage, the optimal level of production is obtained according to the profit maximization of private firms. Based on Eq. (2.10), by solving the profit maximization problem, the optimal level of production for the private firms is obtained in the function of the emission tax rate and the number of the firms, respectively, as follows.

$$q^i = \frac{1-s}{2+n} \tag{2.11}$$

Equation (2.11) demonstrates that, in this case, the equilibrium output level is higher than the case of the linearly proportional emission tax in the previous section (Eq. 2.7), because $s < t$, and it suggests that the output level of each firm is higher in this case.

Profit of each firm is shown by substituting Eq. (2.11) into (2.10) as follows.

$$\Pi^i = \frac{(1-s)\left[(2+n)(1-ns)-(1-s)\left(n+\frac{1}{2}\right)\right]}{(n+2)^2} - F^2 \tag{2.12}$$

In the second stage, we calculate the equilibrium of the industry. Due to the assumption of free entry and withdrawal, the profit of each firm is zero in the Cournot-Nash equilibrium. By setting Eq. (2.12) as equal to zero, we obtain the equilibrium number of the firms under the free-entry market equilibrium as follows.

$$n = \frac{4F^2 + s(1-s) + \sqrt{[s(s-1)-4F^2]^2 - 4[s(s-1)-F^2][\frac{3}{2} - S\left(1+\frac{1}{2}s\right)]}}{2[s(s-1)-F^2]} \tag{2.13}$$

Equation (2.13) implies that the equilibrium number of the firms again depends on the entry cost and the emission tax rate. Furthermore, it suggests that the equilibrium number of the firms decreases (increases) according to the increase (decrease) of the entry cost and the emission tax rate.[3] Furthermore, the equilibrium number of the firms is smaller in this case, because the total liability of the emission tax and the marginal cost is the same as in the previous section.

2.4 Mixed Economy with Public Firm (Model B)

Here, we introduce a public firm, 0, into our model developed in the previous two sections. For analytical simplicity, an entry cost does not apply to the public firm.[4]

[3]We obtained the same level of output, profit of each firm, and the equilibrium number of the firms by setting $s = t/n$ in Eqs. (2.10), (2.11), (2.12), and (2.13).

[4]We assume that the public firm was established before the production of the market begins.

The government holds the public firm, and its objective is to maximize the social welfare by correcting the total amount of output and emission damages. The social welfare consists of the consumer's surplus, the producer's surplus, and the emission tax revenue minus the environmental damages.

$$W = CS + \sum\nolimits_{i=1}^{n} \Pi^i + \Pi^0 + T - ED \tag{2.14}$$

Game procedure also follows three stages. In the first stage, the government sets the emission tax rate $T = t\mu^i = tq^i$ before the entry of the firms according to the social welfare maximization. In the second stage, private firms decide whether they enter the market and produce the goods according to the expected profit given the emission tax rate. Finally, in the third stage, private firms and the public firm set their level of production in the Cournot-Nash equilibrium and are in competition with each other.

The profit function for private firms and the public firm are shown in Eqs. (2.15) and (2.16), respectively.

$$\Pi^i = Pq^i - \frac{\left(q^i\right)^2}{2} - t\mu^i - F^2 \quad (i = 0, 1, \ldots, n) \tag{2.15}$$

$$\Pi^0 = Pq^0 - \frac{\left(q^0\right)^2}{2} - t\mu^0 \tag{2.16}$$

The sub-game perfect Nash equilibrium can again be solved by the recursive induction method as follows.

In the third stage, the optimal level of production is obtained according to the profit maximization of private firms and social welfare maximization of the public firm. Based on Eqs. (2.15) and (2.16), we solve the profit maximization problem; the optimal level of production for the private firms and the public firm is obtained in the emission tax rate function and the number of the firms through Eqs. (2.17) and (2.18), respectively.

$$q^i = \frac{2 - 3t}{6 + n} \tag{2.17}$$

$$q^0 = \frac{2 + n\,(2t - 1)}{6 + n} \tag{2.18}$$

Profit of the private firms and the public firm is found by substituting Eqs. (2.17) and (2.18) into (2.19) and (2.20), respectively.

$$\Pi^i = \frac{3(2 - 3t)^2}{2(6 + n)^2} - F^2 \tag{2.19}$$

$$\Pi^0 = \frac{[2 + n\,(2t - 1)]\,[(6 + n)\,(2t - 1)]}{2(6 + n)^2} \tag{2.20}$$

In the second stage, by setting Eq. (2.19) as equal to zero, we obtain the equilibrium number of the private firms under the equilibrium as follows.

$$n = \frac{\sqrt{6}(2-3t) - 12F}{2F} \tag{2.21}$$

Equation (2.21) implies that the equilibrium number of the firms depends on the entry cost and the emission tax rate and that the equilibrium number of the firms decreases (increases) according to the increase (decrease) of the entry cost and the emission tax rate.

Finally, in the first stage, the government sets the optimal emission tax rate by maximizing the social welfare. As the profit of each firm is zero in the industrial equilibrium, the government sets the rate according to the consumer's surplus maximization. The optimal rate of the emission tax is

$$t = \frac{3 - 2\sqrt{6}F}{6}. \tag{2.22}$$

As the level of the output of public enterprise is larger than that of each private firm, public production has a certain role in alleviating the inefficiency caused by less output than the optimal level in the oligopolistic market. Furthermore, as the equilibrium number of the private firms is smaller than that in Sect. 2.2, public production is also effective in decreasing the excess entry of private firms.[5]

2.5 High-Cost and Cleaner Production Technology and the Role of the Public Firm

In this section, we introduce high-cost production technology and study the inducement policy regarding abatement of pollutant emissions. We assume that the public firm has more advanced production technology that abates pollutant emissions but incurs higher cost. Here, we consider the cost function of the public firm as

$$C\left(q^0\right) = \alpha \cdot \frac{\left(q^0\right)^2}{2} \quad (\alpha > 1) \tag{2.23}$$

Under the cleaner production technology, the emission and the environmental damages of the externality of the public firm are shown as

$$ED = \frac{\left(\sum_{j=0}^{n} \mu^j\right)^2}{2} + \frac{1}{\alpha}\mu^0. \tag{2.24}$$

[5]See Mankiw and Whinston (1986).

The government imposes the emission tax $T = t\mu^i = tq^i$ on the private firms and the public firm.

The profit function of the private firms is assumed to be the same as in the previous three sections. Therefore, the profit function of the public firm in this setting perceives the abovementioned change and it is shown as

$$\Pi^0 = Pq^0 - \alpha \cdot \frac{\left(q^0\right)^2}{2} - t\frac{\mu^{0\,0}}{\alpha} \tag{2.25}$$

As in Sect. 2.3, the social welfare consists of the consumer's surplus, the producer's surplus, and the emission tax revenue minus the environmental damages.

$$W = CS + \sum\nolimits_{i=1}^{n} \Pi^i + \Pi^0 + T - ED \tag{2.26}$$

Using the same method, Eqs. (2.27) and (2.28) show the optimal level of production under the Cournot-Nash equilibrium for the private firms and the public firm, respectively.

$$qi = \frac{(3\alpha - 5)\,t + 2\,(\alpha + 2)}{2\,[n\,(1 + \alpha) + 2\,(3 + \alpha)]} \tag{2.27}$$

$$q^0 = \frac{4 + n\,[(3 - 5\alpha)\,t - 2] - 2\,(5\alpha + 1)\,t}{2\,[n\,(1 + \alpha) + 2\,(3 + \alpha)]} \tag{2.28}$$

To investigate the effects of the induction policy of more advanced technology, we check the changes perceived by the level of production caused by increases in α. By differentiating Eqs. (2.27) and (2.28) with α, we obtain the following two equations.

$$\frac{\partial qi}{\partial \alpha} = \frac{4t\,(14 - n) - 4\,(n - 2)}{4[n\,(1 + \alpha) + 2\,(3 + \alpha)]^2} \tag{2.29}$$

$$\frac{\partial q0}{\partial \alpha} = \frac{n\,(2t - 1) + 3t}{4[n\,(1 + \alpha) + 2\,(3 + \alpha)]^2} \tag{2.30}$$

Equation (2.29) is negative if n is sufficiently large ($n > 14$), and it implies that the level of output of each private firm decreases as α increases. The introduction of higher-cost and cleaner technology in the public firm induces less production by the firms with conventional production technology. However, the introduction of the higher-cost and cleaner technology may increase or decrease the level of public production as Eq. (2.30) can take both signs.

2.6 Conclusion

We studied the environmental policies and sustainability of industry in a free-entry Cournot-Nash oligopoly mixed market with a welfare-maximizing public firm. In addition, we investigated the effects of an induction policy regarding emission abatement technology implemented by the public firm.

We obtained the following findings. The equilibrium level of production for each firm and the number of firms is smaller in the case of total emission tax than the proportional emission tax levied on the level of production. This result suggests that the introduction of the total emission tax may alleviate the inefficiency of the free-entry oligopolistic market. Second, introduction of public production has a certain role in alleviating the inefficiency by increasing the output and decreasing the possible excess entry of the private firms. Lastly, introducing higher-cost and cleaner technology in the public firm may improve efficiency by inducing less production by firms with inferior production technology.

In our future study, we aim to study the effects of inducing policy of more advanced technology into private firms and investigate a possible welfare improvement in a free-entry oligopolistic market with public production.

References

Barnett AH (1980) The Pigouvian tax rule under monopoly. Am Econ Rev 70:1037–1041

Brandao A, Castro S (2007) State owned enterprises as indirect instruments of entry regulation. J Econ 92(3):263–274

Capuano C, De Feo G (2010) Privatization in oligopoly: the impact of the shadow cost of public funds. Rivista Italiana Degli Economisti 15(2):175–208

Cato S, Matsumura T (2015) Optimal privatization and trade policies with endogenous market structure. Econ Rec 91(294):309–323

Damania D (1996) Pollution taxes and pollution abatement in an oligopoly supergame. J Environ Econ Manag 30:323–336

Davis OA, Whinston A (1962) Externalities, welfare and the theory of games. J Polit Econ 70: 241–262

Fujiwara K (2009) Environmental policies in a differentiated oligopoly revisited. Resour Energy Econ 31:239–247

Ino H, Matsumura T (2010) What role should public enterprise play in free entry market. J Econ 101:213–230

Katsoulacos Y, Xepapadeas A (1995) Environmental policy under oligopoly with endogenous market structure. Scand J Econ 97(3):411–420

Lee SH (1999) Optimal taxation for polluting oligopolists with endogenous market structure. J Regul Econ 15(3):293–308

Lili X, Sang-Ho L (2018) Environmental policies with excess burden of taxation in free-entry mixed markets. Int Rev Econ Financ 58:1–13

Mankiw NG, Whinston MD (1986) Free entry and social inefficiency. RAND J Econ 17:48–58

Matsumura T, Kanda O (2005) Mixed oligopoly at free entry markets. J Econ 84:27–48

Pigou AC (1920) The economics of welfare. Macmillan, London

Chapter 3
Emission Taxation and Investment in Cleaner Production: The Case of Differentiated Duopoly

Yasunori Ouchida, Makoto Okamura, and Yukiko Orito

Abstract Ouchida et al. (Cleaner production technology and optimal emission tax, mimeo. 2017) examine the cleaner production technology of the pollution abatement and specify the technology as a log form. They develop the following three-stage game. In the first stage, a government sets a pollution tax rate. In the second stage, duopolistic firms decide its level of abatement investment. In the third stage, the firms engage in Cournot competition in a homogeneous product market. They obtain the explicit solution of the perfect Nash equilibrium of the game. No previous studies have derived the explicit solution of this three-stage game. By incorporating differentiated product markets into the third stage of the game, we generalize Ouchida et al. (Cleaner production technology and optimal emission tax, mimeo. 2017). We derive the explicit equilibrium values of the optimal tax rate, the level of the abatement investment, and the outputs.

Keywords Cleaner production technology · Differentiated duopoly · Optimal emission tax

JEL Classification Q55; Q58; L13; O38.

Y. Ouchida (✉) · Y. Orito
Department of Economics, Hiroshima University, Higashi-Hiroshima, Japan
e-mail: ouchida@hiroshima-u.ac.jp; orito@hiroshima-u.ac.jp

M. Okamura
Economics Department, Gakushuin University, Toshima-ku, Japan

Department of Economics, Hiroshima University, Higashi-Hiroshima, Japan
e-mail: andrew-uf@galaxy.plala.or.jp

K. Nakayama, Y. Miyata (eds.), *Theoretical and Empirical Analysis in Environmental Economics*, New Frontiers in Regional Science: Asian Perspectives 34, https://doi.org/10.1007/978-981-13-2363-8_3

3.1 Introduction

A vast of studies has been accumulated in the field of environment protection, especially environmental abatement technology. The environmental abatement technology is classified into two types: (i) the end-of-pipe technology and (ii) the cleaner production technology (CP technology). The end-of-pipe technology can absorb emissions at the end of the production process. This technology makes the production process, itself, cleaner. On the other hand, the CP technology can modify the production process itself. In other words, the CP technology can decrease the emissions/production ratio in the production process. Typical example of CP technology is flue gas desulfurization equipment and activated carbon absorption equipment. Frondel et al. (2007) cite examples of CP technologies as "recirculation of materials, the use of environmentally friendly materials (e.g., replacing organic solvents by water), and modification of combustion chamber design (process-integrated systems)." A lot of attention has been focused on the CP technology to tackle environmental issues.

Many researchers have developed various types of game models with the end-of-pipe technologies. For example, see Wang and Wang (2009). By contrast, few studies exist on the CP technology, although many firms have adopted the technology. The reason for this research gap is analytical difficulties of the CP technology. That is, a three-stage game model including the CP technology has not been solved analytically because of the tractable difficulties caused by assumed properties of the emission functions and abatement cost functions.[1]

Petrakis and Xepapadeas (1999, 2003), Chiou and Hu (2001), Tsai et al. (2015), and others have developed models that include the CP technology. These papers adopt a quadratic function for environmental investment. However, this quadratic assumption implies the unrealistic property that the emissions/production goes down to zero at the level of limited cost. This zero property is inconsistent with the technological limitation. Considering this shortcoming, Ouchida et al. (2017) provide an alternative form of the environmental abatement function in which the emission/production ratio does not decrease to zero at limited investment level. They address the efficient formation of environmental investment in a homogenous goods Cournot competition. The government can precomit the emission tax rate. They can derive the analytically explicit solution of the following three-stage game: (i) the government sets its emission tax rate. (ii) Two firms decide the environmental abatement investment. (iii) The firms determine the outputs in a homogenous goods market. The purpose of this paper is to generalize Ouchida et al. (2017) by incorporating production differentiation and derive the explicit solution of the game with product differentiation.

The rest of this paper is organized as follows. Section 3.2 formulates a three-stage game model. In Sects. 3.3 and 3.4, we explicitly solve the game under

[1]For example, see Katsoulacos and Xepapadeas (1996), Petrakis and Xepapadeas (1999, 2003, Section 3), Chiou and Hu (2001), and Ben Youssef (2009, 2010, 2011).

non-cooperative environmental investment cases. Section 3.5 presents concluding remarks.

3.2 The Model

Consider a differentiated Cournot duopoly in which a government imposes an emissions tax. The duopolistic firms pollute with emissions and adopt a CP technology. We assume that each firm has an identical environmental technology and cost structure. We develop the following three-stage game.

Stage 1: The government chooses the emissions tax rate on each firm.
Stage 2: Each firm determines environmental investment level non-cooperatively or cooperatively.
Stage 3: Each firm sets its own production level.

Each firm produces a differentiated good. According to Singh and Vives (1984), a representative consumer's utility function is assumed to be

$$U(q_i, q_j) = A(q_i + q_j) - \frac{1}{2}(q_i^2 + 2bq_iq_j + q_j^2) + Y, \tag{3.1}$$

where $A(> 0)$ is the market size parameter and $Y(> 0)$ captures the consumption of a numeraire good. Firm i produces the good i and q_i denotes firm i's production level. Both goods are substitutable; the value of $b(\in [0, 1])$ denotes the parameter of product differentiation. When $b = 1 (b = 0)$, both goods are completely homogeneous (independent). Ouchida et al. (2017) analyze the case with $b = 1$. The demand for good i is

$$p_i(q_i, q_j) = A - q_i - bq_j, \ i, j = 1, 2; i \neq j. \tag{3.2}$$

Consumer surplus CS is given by $CS = (1/2)(q_i^2 + 2bq_iq_j + q_j^2)$.

The production of each good generates emissions. The initial level of the emissions/production ratio is given as $\bar{a}(> 0)$. The government determines the per unit emissions tax rate t to encourage the firm's abatement effort. Each firm has an incentive to invest in cleaner production. Firm i can reduce the level of the emissions/production ratio from $\bar{a}$ to $\bar{a} - a_i$ $(a_i \in [0, \bar{a}))$, through incurring a cost $f(a_i)$, where a_i is a level of effort.

Different from previous literatures, we assume that $f(0) = 0, f' > 0, f'' > 0$, and $\lim_{a_i \to \bar{a}-0} f(a_i) = +\infty$ (see Fig. 3.1).[2] This assumption reflects the reality that existing technology is unable to reduce the value of the emissions/production

[2]Katsoulacos and Xepapadeas (1996, Section 4), Chiou and Hu (2001), Ben Youssef (2009, 2010, 2011), and Tsai et al. (2015) employ a quadratic cost function.

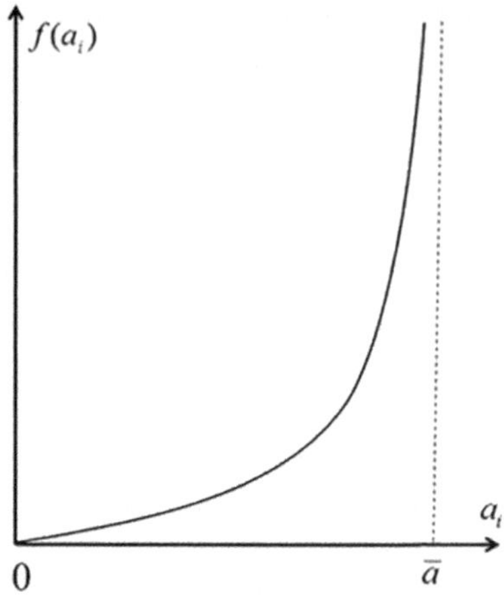

Fig. 3.1 Environmental investment cost

ratio to zero. We specify the functional form $f(a_i)$ as Assumption 1. Under Assumption 1, we can explicitly derive the subgame perfect Nash equilibrium of the above three-stage game. To our knowledge, previous literatures cannot obtain the explicit solution.

Assumption 1 $f(a_i) = -\log(\bar{a} - a_i) + \log \bar{a}$.

The emission function of firm i is specified by

$$e_i \equiv \max\{\bar{a} - a_i, 0\}q_i. \tag{3.3}$$

This function, e_i, is substantially identical to one used in several papers.[3] Total emissions, $E \equiv e_1 + e_2$, result in an environmental damage $D(E) \equiv (d/2)E^2$. The value of $d(> 0)$ shows the damage coefficient, representing consumer's concern for quality of environment.

The total cost function of firm i is assumed to be additively separable with respect to production cost q_i^2 and investment cost $f(a_i)$. Tax payments are given by te_i, where t is an emissions tax rate. Firm i's profit is defined as

$$\pi_i = p_i(q_i, q_j)q_i - q_i^2 - \{-\log(\bar{a} - a_i) + \log \bar{a}\} - te_i. \tag{3.4}$$

Total surplus W is defined as the sum of consumer surplus, producer profits, and the tax revenues, $T \equiv tE$, minus the amount of environmental damage:

$$W \equiv CS + \pi_1 + \pi_2 + T - D(E). \tag{3.5}$$

[3]For example, see Chiou and Hu (2001), Petrakis and Xepapadeas (1999, 2003, Section 3), Ben Youssef (2010), Cato (2011), and Hattori (2013).

The government imposes t to maximize W. We examine two scenarios: (i) non-cooperative and (ii) cooperative environmental investment. In the non-cooperative case, each firm sets investment level to maximize its own profit, while in the cooperative case, each firm doses it to maximize the joint profit.

3.3 The Perfect Equilibrium of the Non-cooperative Environmental Investment: Case (i)

In this section, we calculate the explicit solution of the game analytically. We adopt the subgame perfect Nash equilibrium (SPNE) as solution concept. We assume $d = 1$ and $\bar{a} = 1$.[4] We derive SPNE of the game by usual backward induction.

3.3.1 The Third Stage: Production

Firm i's profit is

$$\pi_i = \{A - q_i - bq_j\}q_i - q_i^2 + \log(1 - a_i) - t(1 - a_i)q_i, \tag{3.6}$$

where $i, j = 1, 2; i \neq j$, for given t. Firm i simultaneously and non-cooperatively determines q_i to maximize its own profit, given the level of a_i, t, and q_j. The first-order condition (FOC) is

$$\frac{\partial \pi_i}{\partial q_i} = A - 4q_i - bq_j - t(1 - a_i) = 0. \tag{3.7}$$

From (3.7), the equilibrium output is written as a function of a_i and a_j.

$$q_i(a_i, a_j) = \frac{(4 - b)A - t[4(1 - a_i) - b(1 - a_j)]}{(4 + b)(4 - b)}. \tag{3.8}$$

3.3.2 The Second Stage: Environmental Investment

Foreseeing the third stage equilibrium (3.8), firm i's profit at stage 2 is given by

$$\pi_i(a_i, a_j) = 2[q_i(a_i, a_j)]^2 + \log(1 - a_i), \ (i, j = 1, 2; i \neq j). \tag{3.9}$$

[4]The assumptions of $d = 1$ and $\bar{a} = 1$ are consistent with Wang and Wang's (2009) model. In addition, $\bar{a} = 1$ is identical to the value used by Ben Youssef (2010), Bárcena-Ruiz and Campo (2012), Ouchida and Goto (2014, 2016a,b), Liu et al. (2015), Pal and Saha (2015), and Moner-Colonques and Rubio (2016).

Firm i simultaneously and non-cooperatively chooses a_i to maximize its own profit, for given a_j. The FOC is the following.[5]

$$\frac{\partial \pi_i}{\partial a_i} = 4q_i(a_i, a_j)\left(\frac{\partial q_i}{\partial a_i}\right) - \frac{1}{1-a_i} = 0. \quad (3.10)$$

We focus on symmetric equilibrium: $a_i = a_j = a$. From (3.10), the equilibrium abatement effort becomes the solution of the following quadratic equation:

$$16t^2a^2 - 16t(2t - A)a + 16t(t - A) + (4 - b)(4 + b)^2 = 0. \quad (3.11)$$

The equilibrium abatement effort is derived as

$$a(t) = \begin{cases} 0 & if \ \text{(i) } X < 0 \text{ or (ii) } t \leq \dfrac{2A - \sqrt{X}}{4}, \ X \geq 0 \\ 1 - \dfrac{2A - \sqrt{X}}{4t} & if \ t > \dfrac{2A - \sqrt{X}}{4}, \text{ and } X \geq 0, \end{cases} \quad (3.12)$$

where $X \equiv 4A^2 - (4 - b)(4 + b)^2$. If $X \geq 0$, then (3.11) has two distinct real roots. However, when $X < 0$, then $a_i = 0$ maximizes firm i's profit. From (3.12), firms carry out no environmental investment when a lower tax rate is precommitted or when the market size, A, is small and the product becomes sufficiently differentiated. After substituting (3.12) into (3.8) and performing some manipulation, we obtain the equilibrium values.[6] The results are presented in Appendix A.

3.3.3 *The First Stage: Emissions Tax*

From (3.20)–(3.25) in Appendix A, total surplus, W, is calculated as a function of t

$$W(t) = \begin{cases} W_1^N(t) = (1+b)\left(\dfrac{A-t}{4+b}\right)^2 + 4\left(\dfrac{A-t}{4+b}\right)^2 + \dfrac{2t(A-t)}{4+b} - 2\left(\dfrac{A-t}{4+b}\right)^2 \\ \qquad if \ X < 0 \text{ or } t \leq \dfrac{2A-\sqrt{X}}{4}, \ (X \geq 0) \\ W_2^N(t) = (1+b)\left(\dfrac{2A+\sqrt{X}}{4(4+b)}\right)^2 + 2\left\{2\left(\dfrac{2A+\sqrt{X}}{4(4+b)}\right)^2 + \log\left(\dfrac{2A-\sqrt{X}}{4t}\right)\right\} \\ \qquad + \dfrac{(4-b)(4+b)}{8} - \dfrac{(4-b)^2(4+b)^2}{128t^2} \quad if \ t > \dfrac{2A-\sqrt{X}}{4}, \ (X \geq 0). \end{cases} \quad (3.13)$$

[5]The second derivatives are $\partial^2\pi_i/\partial a_i^2 = \frac{4t^2}{(4+b)^2(4-b)^2} - \frac{1}{(1-a_i)^2}$. Noting that the value of a_i is in $[0, 1)$, if at least $t < 15/2$, the second-order condition (SOC) is satisfied.

[6]Consequently, none of output, consumer surplus, or total tax revenue depend on t.

where N shows the non-cooperative case. The government determines t to maximize $W(t)$. The FOC is[7]

$$\frac{dW(t)}{dt}=0 \Longleftrightarrow \begin{cases} \dfrac{dW_1^{\mathrm{N}}(t)}{dt}=\dfrac{2[A-t(5+b)]}{(4+b)^2}=0 \quad if \ \text{(i) } X<0 \text{ or (ii) } t\le \dfrac{2A-\sqrt{X}}{4},\ X\ge 0 \\ \dfrac{dW_2^{\mathrm{N}}(t)}{dt}=\dfrac{(4-b)^2(4+b)^2-128t^2}{64t^3}=0 \quad if \ t>\dfrac{2A-\sqrt{X}}{4}, \text{ and } X\ge 0. \end{cases} \tag{3.14}$$

From (3.14), we see that the tax rate $t_{\mathrm{N1}} = A/(5+b)$ maximizes $W_1^{\mathrm{N}}(t)$. Similarly, the tax rate $t_{\mathrm{N2}} = \sqrt{2}(4-b)(4+b)/16$ maximizes $W_2^{\mathrm{N}}(t)$. The government chooses the optimal tax rate by comparing $W_1^{\mathrm{N}}(t_{\mathrm{N1}}) \equiv W_{\mathrm{N1}}$ and $W_2^{\mathrm{N}}(t_{\mathrm{N2}}) \equiv W_{\mathrm{N2}}$. We now define $g(A,b)=0$ as the parameters set of $A>0$ and $b\in[0,1]$, which satisfies $W_{\mathrm{N1}} = W_{\mathrm{N2}}$. The locus $g(A,b)=0$ is depicted in Fig. 3.2. In the left (right) region of $g(A,b)=0$, that is, in Region I (Regions II and III), $W_{\mathrm{N1}} > (<) W_{\mathrm{N2}}$. Lemma 1 summarizes this result.

Lemma 1 *For all $b\in[0,1]$ and $A>0$, a set of product differentiation and demand parameters exists such that*

(i) $W_{\mathrm{N1}} > W_{\mathrm{N2}}$ (Region I).
(ii) $W_{\mathrm{N1}} < W_{\mathrm{N2}}$ (Regions II and III).

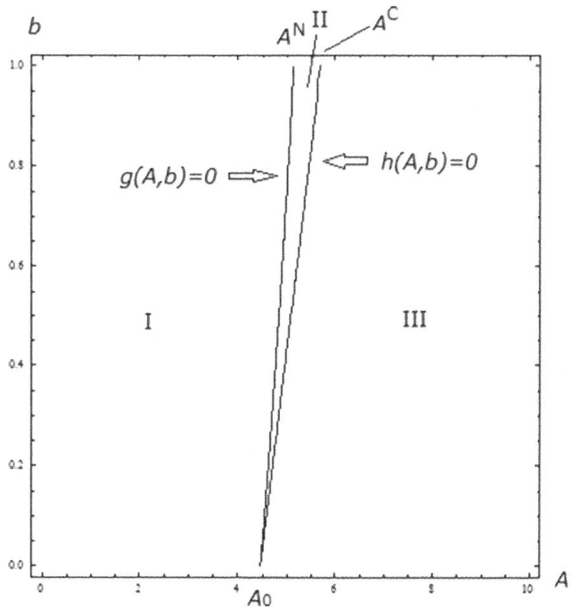

Fig. 3.2 Comparison of total surplus

[7] The SOCs are satisfied because $d^2W_1^{\mathrm{N}}(t)/dt^2<0$ and $d^2W_2^{\mathrm{N}}(t)/dt^2<0$ for all $b\in[0,1]$.

Proof See Fig. 3.2. □

Let us define $A^{\mathrm{N}} \equiv \{A(>0)\,|W_{\mathrm{N1}} = W_{\mathrm{N2}} \text{ and } b = 1\}$. Then, we obtain $A^{\mathrm{N}} \approx 5.1357$. We show Lemma 2.

Lemma 2 *If* $A > A^{\mathrm{N}}$, *then* $W_{\mathrm{N1}} < W_{\mathrm{N2}}$ *holds for all* $b \in [0, 1]$.

Proof See Fig. 3.2. □

In the Region I (Regions II and III), the government precommits and sets t_{N1} (t_{N2}). After some manipulation, we derive the SPNE outcomes as shown in Table 3.1.[8]

3.4 The Perfect Equilibrium of Cooperative Environmental Investment: Case (ii)

3.4.1 The Third Stage: Production

The analysis of stage 3 is identical to that of Sect. 3.3.1.

3.4.2 The Second Stage: Environmental Investment

From (3.9), we have joint profit: $\Pi(a_i, a_j) \equiv \pi_i(a_i, a_j) + \pi_j(a_i, a_j)$. Each firm chooses its own abatement effort level to maximize the joint profit:

$$\Pi(a_i, a_j) = 2[q_i(a_i, a_j)]^2 + \log(1 - a_i) + 2[q_j(a_i, a_j)]^2 + \log(1 - a_j). \quad (3.15)$$

The FOC is[9]

$$\frac{\partial \Pi}{\partial a_i} = 4q_i(a_i, a_j)\left(\frac{\partial q_i}{\partial a_i}\right) - \frac{1}{1 - a_i} + 4q_j(a_i, a_j)\left(\frac{\partial q_j}{\partial a_i}\right) = 0. \quad (3.16)$$

From (3.16), the equilibrium abatement effort is given by

$$a(t) = \begin{cases} 0 & if \ \text{(i) } Z < 0 \text{ or (ii) } t \leq \dfrac{A - \sqrt{Z}}{2},\ Z \geq 0 \\ 1 - \dfrac{A - \sqrt{Z}}{2t} & if \ t > \dfrac{A - \sqrt{Z}}{2}, \text{ and } Z \geq 0, \end{cases} \quad (3.17)$$

[8] Subscript N stands for the non-cooperative environmental investment, and subscripts N1 and N2 express the equilibrium values realized in Region I and Regions (II and III), respectively, in Fig. 3.2. Because $\max\{t_{\mathrm{N1}}, t_{\mathrm{N2}}\} < 15/2$ for all $b \in [0, 1]$, the SOC at stage 2 is satisfied.

[9] The SOC is $\partial^2 \Pi_i / \partial a_i^2 = \frac{4t^2}{(4+b)^2(4-b)^2} - \frac{1}{(1-a_i)^2} + 4[\partial q_j / \partial a_i]^2 < 0$. For details, see footnote 13.

Table 3.1 Equilibrium outcomes under two scenarios

	Two scenarios	
	Non-cooperative environmental investment	Cooperative environmental investment
Emissions tax rate	$t_N = \begin{cases} t_{N1} = \dfrac{A}{5+b} \\ t_{N2} = \dfrac{\sqrt{2}(4-b)(4+b)}{16} \end{cases}$	$t_C = \begin{cases} t_{C1} = \dfrac{A}{5+b} \\ t_{C2} = \dfrac{\sqrt{2}(4+b)}{4} \end{cases}$
Abatement efforts	$a_N = \begin{cases} a_{N1} = 0 \\ a_{N2} = 1 - \dfrac{2A - \sqrt{X}}{4t_{N2}} \end{cases}$	$a_C = \begin{cases} a_{C1} = 0 \\ a_{C2} = 1 - \dfrac{A - \sqrt{Z}}{2t_{C2}} \end{cases}$
Output level	$q_N = \begin{cases} q_{N1} = \dfrac{A}{5+b} \\ q_{N2} = \dfrac{2A + \sqrt{X}}{4(4+b)} \end{cases}$	$q_C = \begin{cases} q_{C1} = \dfrac{A}{5+b} \\ q_{C2} = \dfrac{A + \sqrt{Z}}{2(4+b)} \end{cases}$
Firm's emission	$e_N = \begin{cases} e_{N1} = q_{N1} \\ e_{N2} = \sqrt{2}/2 \end{cases}$	$e_C = \begin{cases} e_{C1} = q_{C1} \\ e_{C2} = \sqrt{2}/2 \end{cases}$
Total emissions	$E_N = \begin{cases} E_{N1} = 2q_{N1} \\ E_{N2} = \sqrt{2} \end{cases}$	$E_C = \begin{cases} E_{C1} = 2q_{C1} \\ E_{C2} = \sqrt{2} \end{cases}$
Environmental damage	$D_N = \begin{cases} D_{N1} = 2[q_{N1}]^2 \\ D_{N2} = 1 \end{cases}$	$D_C = \begin{cases} D_{C1} = 2[q_{C1}]^2 \\ D_{C2} = 1 \end{cases}$
Profit	$\pi_N = \begin{cases} \pi_{N1} = [A - (1+b)q_{N1}]q_{N1} - [q_{N1}]^2 - t_{N1}q_{N1} \\ \pi_{N2} = [A - (1+b)q_{N2}]q_{N2} - [q_{N2}]^2 + \log(1 - a_{N2}) - t_{N2}[1 - a_{N2}]q_{N2} \end{cases}$	$\pi_C = \begin{cases} \pi_{C1} = [A - (1+b)q_{C1}]q_{C1} - [q_{C1}]^2 - t_{C1}q_{C1} \\ \pi_{C2} = [A - (1+b)q_{C2}]q_{C2} - [q_{C2}]^2 + \log(1 - a_{C2}) - t_{C2}[1 - a_{C2}]q_{C2} \end{cases}$
Consumer surplus	$CS_N = \begin{cases} CS_{N1} = (1+b)[q_{N1}]^2 \\ CS_{N2} = (1+b)[q_{N2}]^2 \end{cases}$	$CS_C = \begin{cases} CS_{C1} = (1+b)[q_{C1}]^2 \\ CS_{C2} = (1+b)[q_{C2}]^2 \end{cases}$
Total surplus	$W_N = \begin{cases} W_{N1} = CS_{N1} + 2\pi_{N1} + t_{N1}E_{N1} - D_{N1} \\ W_{N2} = CS_{N2} + 2\pi_{N2} + t_{N2}E_{N2} - D_{N2} \end{cases}$	$W_C = \begin{cases} W_{C1} = CS_{C1} + 2\pi_{C1} + t_{C1}E_{C1} - D_{C1} \\ W_{C2} = CS_{C2} + 2\pi_{C2} + t_{C2}E_{C2} - D_{C2} \end{cases}$

where $Z \equiv A^2-(4+b)^2$. If $Z \geq 0$, then (3.16) has two distinct real roots. However, when $Z < 0$, then $a_i = 0,\ (i = 1, 2)$ maximizes $\Pi(a_i, a_j)$. From (3.8), $\partial q_j/\partial a_i < 0$ holds. Therefore, (3.16) implies that the equilibrium abatement effort level under investment coordination, $1 - [(2A - \sqrt{X})/4t]$, is lower than the non-cooperative case, $1 - [(A - \sqrt{Z})/2t]$.[10] Other results are presented in Appendix B.[11]

3.4.3 The First Stage: Emissions Tax

From (3.26)–(3.31) in Appendix B, total surplus is derived as

$$W(t) = \begin{cases} W_1^C(t) = (1+b)\left(\dfrac{A-t}{4+b}\right)^2 + 4\left(\dfrac{A-t}{4+b}\right)^2 + \dfrac{2t(A-t)}{4+b} - 2\left(\dfrac{A-t}{4+b}\right)^2 \\ \qquad if\ \text{(i) } Z < 0 \text{ or (ii) } t \leq \dfrac{A-\sqrt{Z}}{2},\ Z \geq 0 \\ W_2^C(t) = (1+b)\left(\dfrac{A+\sqrt{Z}}{2(4+b)}\right)^2 + 2\left\{2\left(\dfrac{A+\sqrt{Z}}{2(4+b)}\right)^2 + \log\left(\dfrac{A-\sqrt{Z}}{2t}\right)\right\} \\ \quad + \dfrac{(4+b)}{2} - \dfrac{(4+b)^2}{8t^2} \qquad if\ t > \dfrac{A-\sqrt{Z}}{2}, \text{ and } Z \geq 0. \end{cases} \tag{3.18}$$

The government chooses t to maximize $W(t)$. The corresponding FOC is[12]

$$\frac{dW(t)}{dt} = 0 \iff \begin{cases} \dfrac{dW_1^C(t)}{dt} = \dfrac{2[A-t(5+b)]}{(4+b)^2} = 0 \\ \qquad if\ Z < 0 \text{ or } t \leq \dfrac{A-\sqrt{Z}}{2},\ (Z \geq 0) \\ \dfrac{dW_2^C(t)}{dt} = \dfrac{(4+b)^2 - 8t^2}{4t^3} = 0 \\ \qquad if\ t > \dfrac{A-\sqrt{Z}}{2},\ (Z \geq 0). \end{cases} \tag{3.19}$$

The equilibrium tax rate t_C and other SPNE outcomes are derived as shown in Table 3.1.[13] From (3.19), we see that the tax rate $t_{C1} = A/(5+b)$ maximizes $W_1^C(t)$. Similarly, the tax rate $t_{C2} = \sqrt{2}(4+b)/4$ maximizes $W_2^C(t)$. The government

[10] In fact, $(1 - [(2A - \sqrt{X})/4t]) - (1 - [(A - \sqrt{Z})/2t]) = (\sqrt{X} - 2\sqrt{Z})/4t > 0$.

[11] None of output, consumer surplus, or tax revenue depends on t.

[12] The SOC is satisfied.

[13] Subscript C represents cooperative environmental investment, and subscripts C1 and C2 the equilibrium values realized in Regions (I and II) and Region III, respectively, in Fig. 3.2. Under precommitment of $t_C \in \{t_{C1}, t_{C2}\}$, the SOC at stage 2 is satisfied.

determines the optimal tax rate by comparing $W_1^{\mathrm{C}}(t_{\mathrm{C1}}) \equiv W_{\mathrm{C1}}$ with $W_2^{\mathrm{C}}(t_{\mathrm{C2}}) \equiv W_{\mathrm{C2}}$. We define $h(A, b) = 0$ as the parameters set of $A > 0$ and $b \in [0, 1]$, which satisfies $W_{\mathrm{C1}} = W_{\mathrm{C2}}$. The curve $h(A, b) = 0$ is depicted in Fig. 3.2. In the left (right) region of $h(A, b) = 0$, that is, Regions I and II (Region III), $W_{\mathrm{C1}} > (<) W_{\mathrm{C2}}$ holds. Consequently, we show Lemma 3.

Lemma 3 *For all $b \in [0, 1]$ and $A > 0$, a set of product differentiation and the demand parameters exists such that*

(i) $W_{\mathrm{C1}} > W_{\mathrm{C2}}$ (Regions I and II).
(ii) $W_{\mathrm{C1}} < W_{\mathrm{C2}}$ (Region III).

Proof see Fig. 3.2. □

When we define $A^{\mathrm{C}} \equiv \{A(> 0) \,|W_{\mathrm{C1}} = W_{\mathrm{C2}}$ and $b = 1\}$, we obtain $A^{\mathrm{C}} \approx 5.6947$. Thus, we have Lemma 4.

Lemma 4 *If $A > A^{\mathrm{C}}$, then $W_{\mathrm{C1}} < W_{\mathrm{C2}}$ holds for all $b \in [0, 1]$.*

Proof See Fig. 3.2. □

3.5 Concluding Remarks

We generalize Ouchida et al. (2017) to the game model with the CP technology when the duopolistic firm produces the differentiated goods. We assume that the CP technology takes a log form same as (Ouchida et al. 2017). We develop the three-stage game. In the first stage, the regulator sets the emission tax rate against the duopolistic firm. In the second stage, each firm invests in the environmental abatement technology. In the third stage, each firm decides its output level in the differentiated product market. We obtain the subgame perfect Nash equilibrium of the game. We show that the main results of Ouchida et al. (2017) carry over to our generalized framework.

Appendices

Appendices provide supplementary explanations of Sects. 3.3.2 and 3.4.2.

Appendix A

The following results show the subgame equilibrium of output, consumer surplus, profit, total emissions, total tax revenue, and environmental damage.

$$q(t) = \begin{cases} \dfrac{A-t}{4+b} & if\ \ X < 0 \text{ or } t \le \dfrac{2A-\sqrt{X}}{4},\ (X \ge 0) \\ \dfrac{2A+\sqrt{X}}{4(4+b)} & if\ \ t > \dfrac{2A-\sqrt{X}}{4},\ (X \ge 0), \end{cases} \tag{3.20}$$

$$CS(t) = \begin{cases} (1+b)\left(\dfrac{A-t}{4+b}\right)^2 & if\ \ X < 0 \text{ or } t \le \dfrac{2A-\sqrt{X}}{4},\ (X \ge 0) \\ (1+b)\left(\dfrac{2A+\sqrt{X}}{4(4+b)}\right)^2 & if\ \ t > \dfrac{2A-\sqrt{X}}{4},\ (X \ge 0), \end{cases} \tag{3.21}$$

$$\pi(t) = \begin{cases} 2\left(\dfrac{A-t}{4+b}\right)^2 & if\ \ X < 0 \text{ or } t \le \dfrac{2A-\sqrt{X}}{4},\ (X \ge 0) \\ 2\left(\dfrac{2A+\sqrt{X}}{4(4+b)}\right)^2 + \log\left(\dfrac{2A-\sqrt{X}}{4t}\right) & if\ \ t > \dfrac{2A-\sqrt{X}}{4},\ (X \ge 0), \end{cases} \tag{3.22}$$

$$E(t) = \begin{cases} \dfrac{2(A-t)}{4+b} & if\ \ X < 0 \text{ or } t \le \dfrac{2A-\sqrt{X}}{4},\ (X \ge 0) \\ \dfrac{(4-b)(4+b)}{8t} & if\ \ t > \dfrac{2A-\sqrt{X}}{4},\ (X \ge 0), \end{cases} \tag{3.23}$$

$$T(t) = tE(t) = \begin{cases} \dfrac{2t(A-t)}{4+b} & if\ \ X < 0 \text{ or } t \le \dfrac{2A-\sqrt{X}}{4},\ (X \ge 0) \\ \dfrac{(4-b)(4+b)}{8} & if\ \ t > \dfrac{2A-\sqrt{X}}{4},\ (X \ge 0), \end{cases} \tag{3.24}$$

$$D(E(t)) = \begin{cases} 2\left(\dfrac{A-t}{4+b}\right)^2 & if\ \ X < 0 \text{ or } t \le \dfrac{2A-\sqrt{X}}{4},\ (X \ge 0) \\ \dfrac{(4-b)^2(4+b)^2}{128t^2} & if\ \ t > \dfrac{2A-\sqrt{X}}{4},\ (X \ge 0). \end{cases} \tag{3.25}$$

Appendix B

The following results denote the subgame equilibrium of output, consumer surplus, profit, total emissions, total tax revenue, and environmental damage.

$$q(t) = \begin{cases} \dfrac{A-t}{4+b} & if\ \ Z < 0 \text{ or } t \leq \dfrac{A-\sqrt{Z}}{2},\ (Z \geq 0) \\ \dfrac{A+\sqrt{Z}}{2(4+b)} & if\ \ t > \dfrac{A-\sqrt{Z}}{2},\ (Z \geq 0), \end{cases} \tag{3.26}$$

$$CS(t) = \begin{cases} (1+b)\left(\dfrac{A-t}{4+b}\right)^2 & if\ \ Z < 0 \text{ or } t \leq \dfrac{A-\sqrt{Z}}{2},\ (Z \geq 0) \\ (1+b)\left(\dfrac{A+\sqrt{Z}}{2(4+b)}\right)^2 & if\ \ t > \dfrac{A-\sqrt{Z}}{2},\ (Z \geq 0), \end{cases} \tag{3.27}$$

$$\pi(t) = \begin{cases} 2\left(\dfrac{A-t}{4+b}\right)^2 & if\ \ Z < 0 \text{ or } t \leq \dfrac{A-\sqrt{Z}}{2},\ (Z \geq 0) \\ 2\left(\dfrac{2A+\sqrt{Z}}{2(4+b)}\right)^2 + \log\left(\dfrac{A-\sqrt{Z}}{2t}\right) & if\ \ t > \dfrac{A-\sqrt{Z}}{2},\ (Z \geq 0), \end{cases} \tag{3.28}$$

$$E(t) = \begin{cases} \dfrac{2(A-t)}{4+b} & if\ \ Z < 0 \text{ or } t \leq \dfrac{A-\sqrt{Z}}{2},\ (Z \geq 0) \\ \dfrac{(4+b)}{2t} & if\ \ t > \dfrac{A-\sqrt{Z}}{2},\ (Z \geq 0), \end{cases} \tag{3.29}$$

$$T(t) = tE(t) = \begin{cases} \dfrac{2t(A-t)}{4+b} & if\ \ Z < 0 \text{ or } t \leq \dfrac{A-\sqrt{Z}}{2},\ (Z \geq 0) \\ \dfrac{(4+b)}{2} & if\ \ t > \dfrac{A-\sqrt{Z}}{2},\ (Z \geq 0), \end{cases} \tag{3.30}$$

$$D(E(t)) = \begin{cases} 2\left(\dfrac{A-t}{4+b}\right)^2 & if\ \ Z < 0 \text{ or } t \leq \dfrac{A-\sqrt{Z}}{2},\ (Z \geq 0) \\ \dfrac{(4+b)^2}{8t^2} & if\ \ t > \dfrac{A-\sqrt{Z}}{2},\ (Z \geq 0). \end{cases} \tag{3.31}$$

References

Bárcena-Ruiz JC, Campo ML (2012) Partial cross-ownership and strategic environmental policy. Resour Energy Econ 34:198–210

Ben Youssef S (2009) Transboundary pollution, R&D spillovers and international trade. Ann Reg Sci 43(1):235–250

Ben Youssef S (2010) R&D in cleaner technology and international trade. Int Game Theory Rev 12(1):61–73

Ben Youssef S (2011) Transboundary pollution and absorptive capacity. Environ Model Assess 16(2):205–211

Cato S (2011) Environmental policy in a mixed market: abatement subsidies and emission taxes. Environ Econ Policy Stud 13:283–301

Chiou JR, Hu JL (2001) Environmental research joint ventures under emission taxes. Environ Resour Econ 20:129–146

Frondel M, Horbach J, Rennings K (2007). End-of-pipe or cleaner production? An empirical comparison of environmental innovation decisions across OECD countries. Bus Strateg Environ 16(8):571–584

Hattori K (2013) Environmental innovation and policy harmonization in international oligopoly. Environ Develop Econ 18:162–183

Katsoulacos Y, Xepapadeas A (1996) Environmental R&D, spillovers and optimal policy schemes under oligopoly. In: Xepapadeas A (ed) Economic policy for the environment and natural resources. Edward Elgar, Cheltenham, pp 59–79

Liu CC, Wang LFS, Lee SH (2015) Strategic environmental corporate social responsibility in a differentiated duopoly market. Econ Lett 129:108–111

Moner-Colonques R, Rubio SJ (2016) The strategic use of innovation to influence environmental policy: taxes versus standards. B.E. J Econ Anal Policy 16(2):973–1000

Ouchida Y, Goto D (2014) Do emission subsidies reduce emission? In the context of environmental R&D organization. Econ Model 36:511–516

Ouchida Y, Goto D (2016a) Cournot duopoly and environmental R&D under regulator's precommitment to an emissions tax. Appl Econ Lett 23(5):324–331

Ouchida Y, Goto D (2016b) Environmental research joint ventures and time-consistent emission tax: endogenous choice of R&D formation. Econ Modell 55:179–188

Ouchida Y, Orito Y, Okamura M (2017) Cleaner production technology and optimal emission tax, mimeo

Pal R, Saha B (2015) Pollution tax, partial privatization and environment. Resour Energy Econ 40:19–35

Petrakis E, Xepapadeas A (1999) Does government precommitment promote environmental innovation? In: Petrakis E, Sartzetakis ES, Xepapadeas A (eds), Environmental regulation and market power: competition, time consistency and international trade. Edward Elgar, Cheltenham, pp 145–161

Petrakis E, Xepapadeas A (2003) Location decisions of a polluting firm and the time consistency of environmental policy. Resour Energy Econ 25:197–214

Singh N, Vives X (1984) Price and quantity competition in a differentiated duopoly. RAND J Econ 15(4):546–554

Tsai TH, Tu K, Chiou JR (2015) Tariffs and environmental taxes in the presence of environmental R&D. Environ Resour Econ 60:413–431

Wang LFS, Wang J (2009) Environmental taxes in a differentiated mixed duopoly. Econ Syst 33(4):389–396

Chapter 4
Effects of Environmental Taxes on Forest Conservation: Case of the Water Resources Conservation Fund in Toyota City

Keiko Nakayama, Mastoshi Shirai, and Mitsuo Yamada

Abstract Seventy percent of our country' s land is occupied by forest. Functions of forest cover not only CO_2 reduction but also water source cultivation, conservation of biodiversity, prevention of sediment-related disasters, and so on. Forest conservation is an urgent policy issue for Japan.

Aichi Prefecture Toyota City was the first municipality to implement the forest reservation policy. Since 1994, Toyota City runs a tap water conservation fund to accumulate 1 yen per 1 m^3 of water usage. In 2003, Kochi Prefecture, followed by other prefectures, established a forest environmental tax as a poll tax. Furthermore, the national government is similarly planning to introduce a forest environment tax in 2024.

Preceding studies on forest environmental taxes have so far mostly focused on the background of the policies, survey on actual condition, and comparisons with overseas cases. Therefore, in this paper, we consider Toyota City's taxation elaboration and the forest environmental tax based on efficiency in resource allocation.

Keywords Environmental tax · Forest conservation policy · Cultivation investment · Toyota water resources conservation fund

4.1 Introduction

Forest resources once played a major role in purifying the global environment by absorbing CO_2 and supplying oxygen. However, global economic growth in the twentieth century over the world has depleted forest resources and caused

K. Nakayama (✉) · M. Shirai · M. Yamada
Chukyo University, Nagoya, Japan
e-mail: nakayama@mecl.chukyo-u.ac.jp; msts@mecl.chukyo-u.ac.jp; yamada@mecl.chukyo-u.ac.jp

K. Nakayama, Y. Miyata (eds.), *Theoretical and Empirical Analysis in Environmental Economics*, New Frontiers in Regional Science: Asian Perspectives 34, https://doi.org/10.1007/978-981-13-2363-8_4

further deterioration of the environment. Forest also affects water purification and presents sediment discharge. However, since forest stock is characterized by a public goods property, individuals usually have low interest in forest recharge. Since the maintenance and conservation of the environment through private forest projects have already reached an impossible stage to attain the goals, the government must now design appropriate conservation policies. Toyota City first introduced a forest recharge policy in 1994. It established a water conservation fund to accumulate 1 yen per 1 m^3 of water usage. This fund was collected for forest conservation in upstream areas to form the source of water supply in Toyota City. Toyota City's policy attracted the attention of several municipalities. They followed Toyota policy. Since Kochi Prefecture founded the "forest environment tax" to maintain and promote the public function of forests in 2003, many other prefectures have introduced similar taxes. These taxes are lump-sum taxes, and the national government plans to introduce the forest environmental tax collected as a lump-sum tax (poll tax) in 2024.

Forest preservation charging systems, such as the Japanese Forest Environment Tax and the Water Source Conservation Fund of Toyota City, are rare in other countries. In this paper, we consider Toyota City's forest environment tax and the lump-sum tax method specific to Japan based on efficiency in long-run resource allocation. The structure of this paper is as follows. In Sect. 4.2, we outline the policy of Toyota City Water Supply Conservation Fund. In Sects. 4.3 and 4.4, we construct a theoretical model that explicitly specifies water supply and forest conservation. We discuss the economic efficiency of the lump-sum tax method in Sect. 4.5 and that of the Toyota City Water Supply Conservation Fund Project in Sect. 4.6.

4.2 Toyota Water Supply Water Conservation Project

Toyota City, located approximately in the middle of Aichi Prefecture, is globally known as the base of *Toyota Motor Corporation*. Yahagi River, a *first-class* river that originates from the Southern Alps and flows through the city's northern and southern regions, brings the source of nature in this area. Nearly 70% of Toyota City areas are occupied by farmland and forest.

The history of Toyota City is short. The Toyota Municipal Government was established in 1951. In 1960, Toyota City became a sister city of Detroit, Michigan, United States. Then, with rapid progress owing to the automobile industry, it became the first core city in Aichi Prefecture in 1998 (Fig. 4.1).

The length of the Yahagi River, flowing through central Toyota City, is over 117 km. It has contributed to the revitalization of this area as a mother river that has fostered life since prehistoric times and, with respect to logistics of ship transport, agricultural water, industrial water, and so on in modern times. The Yahagi River originates from Nagano Prefecture from a mountainous source that spans three prefectures: Nagano, Gifu, and Aichi.

Fig. 4.1 Location of Toyota city

Toyota City is located in the middle basin of the Yahagi River, which flows from the Southern Alps into Mikawa Bay, at the point of contact between the mountains and the plains. The upstream catchment area of the Yahagi River reaches 114,000 ha, it's 87% is occupied by the forest. Moreover 58,000 ha of the forest is an artificial forest. Toyota City's water supply system uses more than 70% of its raw water from Yahagi River and the Yahagi Dam. However, forestry has declined markedly owing to aging and depopulation. Timber prices have also been sluggish owing to large imports of foreign timber.

After the Second World War, tree planting was carried out in the catchment area of the Yahagi River. In spite of postwar afforestation in this catchment area, more than 50% of the plantation here had been devastated even at the time of thinning, with devastation being conspicuous around the basin.

In addition to timber production, the forests have many public functions such as water source recharge, water purification, prevention of land disasters, and preservation of the natural environment. However, the decline of forestry has rapidly reduced these functions. It is difficult to seek to forestry stakeholders and residents of areas in order to maintain and restore the public function of the forest. Thus, residents, municipalities, and companies located near downstream water resources that benefit from the public function of forests should work together for forest conservation.

In the Yahagi River basin, basin residents have been attempting to conserve water and clean rivers, based on the belief that the "basin is one community with a common destiny".

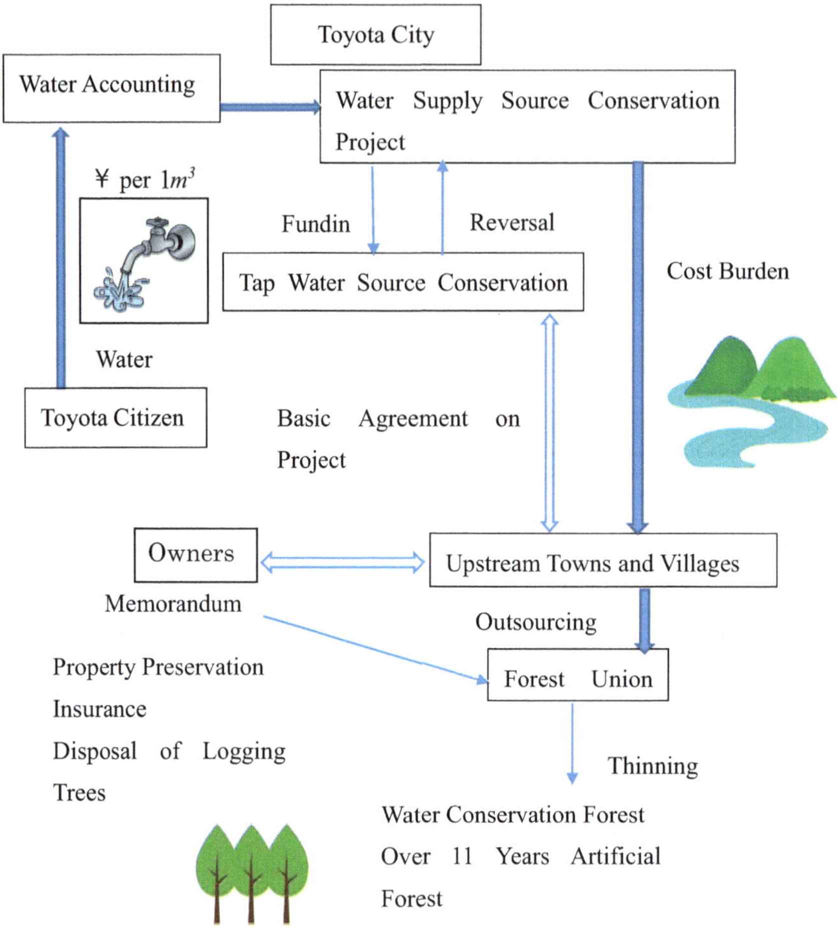

Fig. 4.2 Toyota system

The "Toyota City Water Supply Conservation Fund" was established in Toyota City, and it became the first nationwide fund supported by a water fee of 1 yen per 1 m^3 of water usage.

In 1994, the funds began to be accumulated and lead to thinning projects in 2000. The fund was first intended to preserve forests. Its objective was to ensure that Toyota citizens would bear the cost of fund to express their gratitude to water source areas. In addition, 1 yen per 1 m^3 of water from the collected water fee is paid out as the "water source conservation project cost" from the water business accounting to the water service special account. Figure 4.2 shows the outline of the tap water source conservation project.

Table 4.1 Balance of Toyota city water source conservation fund

FY	Annual income (¥)	Project cost (¥)	Fund's balance (¥)
1994	39,281,014	0	39,281,014
1995	45,056,738	0	84,337,752
1996	45,600,000	0	129,937,752
1997	47,200,000	0	177,137,752
1998	47,250,000	0	224,387,752
1999	48,018,000	0	272,405,752
2000	46,956,236	19,365,153	299,907,000
2001	46,924,000	22,108,802	324,126,000
2002	44,253,165	29,634,056	338,400,000
2003	43,219,898	28,865,829	351,700,000
2004	44,498,975	24,880,881	370,625,000
2005	45,377,648	16,529,531	397,784,000
2006	51,040,534	28,559,797	419,083,000
2007	52,315,708	4,260,000	464,209,000
2008	125,050,114	120,269,697	398,007,815
2009	48,431,421	5,320,000	439,187,815
2010	50,018,164	2,915,000	484,301,815
2011	48,694,525	302,185	530,836,815
2012	49,051,039	8,124,470	570,559,815
2013	48,282,065	8,581,500	608,989,315
2014	48,753,875	17,998,383	638,135,315
2015	57,840,291	55,079,893	630,895,783
2016	64,094,579	60,009,924	614,189,889
2017	49,143,565	42,216,104	620,236,889

Toyota City first selects water conservation forests from privately planted forests that have been devastated in the upper stream of the Yahagi River. They are under public management (mainly thinning) for 20 years. The management costs are covered by the special account of the fund. The owner of the forest is forbidden from doing all activities within this period as compensation for public management. Table 4.1 shows the fund status since its establishment.

This fund attracted immense national attention at the time of establishment, and the Aichi Chubu Water Supply Authority established the "Water Supply Environmental Conservation Fund" in 2001 in the same way as Toyota City. Then, the Union of Kiso Governments established the "Kiso Forest Conservation Fund" in 2004. Mizukubo Town, Shizuoka Prefecture (merged with Hamamatsu City, Shizuoka Prefecture in 2005); Fukuoka City, Fukuoka Prefecture, in 1997; Ena City, Gifu Prefecture, in 1999; and Gamagori City, Aichi Prefecture, in 2003 established similar funds. Asuke Town, which was merged into Toyota City, also introduced the "Forest Fund".

The "Toyota City Water Supply Conservation Fund" is a pioneer institution in Japan. It is a worthy attempt toward conservation. The fund was neither established as an obligation for residents of areas near the downstream, nor as a beneficiary burden, but it was created in gratitude. However, all the municipalities (Fujioka Town, Obara Village, Asuke Town, Shimoyama Village, Asahi Town, and Inabu Town) that had received the fund before merging with Toyota City in 2000 must contribute the fund after the merge with Toyota City.

Kochi Prefecture was the first to impose a forest environmental tax. Now, about 80% prefectures in Japan have levied the kind of forest environmental tax. The range of the tax is mostly from 500 yen to 1000 yen per year for individuals, and they are collected by prefectural taxes.

National forest environmental tax is expected to be introduced in 2024. The tax would be levied on the inhabitants of municipalities by 1000 yen per year, and the tax revenue would be used mainly for thinning of untreated private artificial forests.

In the economic analysis of water supply, many researches deal with the problems on efficiency and regulation of water supply and forest conservation. Almost economic researches of forest concern on achieving maximum sustainable output, since forest conservation should be tackled at a global scale.

Hall (2017) considered issues of forest tax, and Gong and Löfgren (2013) assessed tax effectiveness of the management of forest resources. Amacher and Brazee (2013) examined how government's preferences affect the choice of taxes. Other scholars have analyzed the influence of taxes for forest recharge on owners – for example, Siegel (2010) and Kimbell et al. (2010). Cushing and Newman (2018) examined how some kinds of taxes imposed on nonindustrial private forests affect profitability. Stavins and Richards (2005) analyzed the costs of forest-based carbon sequestration, while Liu and Wu (2017) analyzed the forest tax policy using the CGE model. In addition, Choo et al. (2017) examined the viability of offering landowner's property tax subsidies for forest carbon sequestration (called "tax-based subsidy approach"). Japanese forest environmental taxes are less common in other countries.

The Toyota City Water Supply conservation Fund is studied by Kamiya (2000); Honda (2007); Nakayama et al. (2016) and so on.

4.3 Water Supply and Forest Conservancy Model

In the following sections, we construct a simple model of water supply and forest conservancy.

For simplicity, we consider an economy comprising a constant number of families and each family lives for an infinite period. We assume that they consume one kind of good and tap water in each period. We assume that all families have the same utility function:

$$U\left(c_t, W_t\right) \tag{4.1}$$

where c_t is the amount of consumption by the family in period t and W_t is the amount of water consumed by the family in period t. We assume that $U()$ is quasi-concave.

The local government supplies water. The cost function of the water supply is assumed to be composed of the constant marginal costs m and the fixed cost K. We suppose that the forest stock in the water source area decreases the costs of water supply owing to external effects and that one unit of the forest stock decreases the cost of the water supply by n.

We define the cost function of water supply C as

$$C\left(W_t, x_t\right) = K + mW_t - nx_t, \tag{4.2}$$

where W_t is the amount of water supply, x_t is the forest stock, K is the fixed costs in the period t, m is the marginal costs of water supply, and n is the marginal external benefits of forest stock.

We assume that the forest stock is destructed at the rate of $\delta \in [0, 1]$ but grows by cultivating investment I_t. Let $f(I_t)$ be the function of an increase in forest stock by cultivating investment in period t. We denote the increases in the forest stock at period t as $\dot{x}_t$, which is represented as

$$\dot{x}_t = f\left(I_t\right) - \delta x_t \tag{4.3}$$

We suppose that $f'\left(I_t\right) > 0, f''\left(I_t\right) < 0$.

The optimum problem of the families is formalized as maximizing problem of the sum of the present value of all families' utility by allocating their given income between consumption, water expenditure, and investments for the forest cultivation investment.

Then, the problem is formulated as

$$\max_{W,I} \int_0^\infty U\left(c_t, W_t\right) e^{-\rho t} \mathrm{d}t$$

subject to

$$Y = c_t + C\left(W_t, x_t\right) + I_t$$

$$C\left(W_t, x_t\right) = K + mW_t - nx_t$$

$$\dot{x}_t = f\left(I_t\right) - \delta x_t$$

$$x_0 = \overline{x}.$$

$\rho \in [0, 1]$ is the social discount rate and, x_0 is the initial ($t = 0$) forest stock.

We define the present value Hamiltonian, H, as

$$H = U\left(Y - K - mW_t + nx_t - I_t, W_t\right) + \varphi\left\{f\left(I_t\right) - \delta x_t\right\} \tag{4.4}$$

where ϕ is the auxiliary variable.

The first order conditions for maximization are

$$H_W = 0 \tag{4.5}$$

$$H_I = 0 \tag{4.6}$$

$$\dot{\varphi} = \rho\varphi - H_x \tag{4.7}$$

Considering that $H_{WW} < 0$is equal to $m^2U_{cc}^2 - 2mU_{cW} + U_{WW} < 0$, the second order conditions are

$$H_{WW} < 0 \tag{4.8}$$

$$H_{II} > 0 \tag{4.9}$$

We assume that second order conditions are met. Then, from Eqs. (4.5), (4.6) and (4.7), we obtain

$$-U_c m + U_W = 0 \tag{4.10}$$

$$-U_c + \varphi f' = 0 \tag{4.11}$$

$$\dot{\varphi} = \varphi\left(\rho + \delta\right) - U_c n \tag{4.12}$$

Furthermore, from Eqs. (4.10) and (4.11), we obtain

$$mU_c = U_W \tag{4.13}$$

$$U_c = \varphi f' \tag{4.14}$$

Substituting Eq. (4.14) into Eq. (4.12), we obtain

$$\dot{\varphi} = \varphi\left(\rho + \delta - n f'\right) \tag{4.15}$$

In the steady state $\left(\dot{\phi} = 0\right)$, from Eq. (4.15), we obtain

$$nf' = \rho + \delta \tag{4.16}$$

In Eq. (4.16), in the steady state, the recharge expenditure level should be determined so that the productivity measured by the marginal decrease in the water production cost of forest recharge expenditure is equal to the net social discount rate.

4.4 The Optimal Path

The optimal path is represented by Eqs. (4.10), (4.11), (4.3), and (4.15). From Eqs. (4.10) and (4.11), the optimal *W*and *I* are given as the function of *x* and φ, respectively.

$$W = W(x, \varphi) \tag{4.17}$$

$$I = I(x, \varphi) \tag{4.18}$$

By totally differentiating Eqs. (4.10) and (4.11), we obtain

$$\left(m^2U_{cc} - 2mU_{cW} + U_{WW}\right) dW + (mU_{cc} - U_{Wc})\, dI = n\,(mU_{cc} - U_{Wc})\, dx \tag{4.19}$$

$$(mU_{cc} - U_{cW})\, dW + \left(U_{cc} + \varphi f''\right) dI = nU_{cc}dx - f' d\varphi \tag{4.20}$$

By displaying these in the form of a matrix, we obtain

$$\begin{pmatrix} m^2U_{cc} - 2mU_{cW} + U_{WW} & mU_{cc} - U_{cW} \\ mU_{cc} - U_{cW} & U_{cc} + \varphi f'' \end{pmatrix} \begin{pmatrix} dW \\ dI \end{pmatrix} = \begin{pmatrix} n\,(mU_{cc} - U_{cW})\, dx \\ nU_{cc}dx - f' d\varphi \end{pmatrix} \tag{4.21}$$

From Eq. (4.21), the following two equations are derived.

$$\frac{\partial I}{\partial x} = \frac{\begin{vmatrix} m^2U_{cc} - 2mU_{cW} + U_{WW} & n\left(mU_{cc} - U_{cW}\right) \\ mU_{cc} - U_{cW} & nU_{cc} \end{vmatrix}}{|A|} = \frac{n}{|A|}\left(U_{cc}U_{WW} - {U_{cW}}^2\right) \tag{4.22}$$

$$\frac{\partial I}{\partial \varphi} = \frac{\begin{vmatrix} m^2U_{cc} - 2mU_{cW} + U_{WW} & 0 \\ mU_{cc} - U_{cW} & f' \end{vmatrix}}{|A|} = \frac{1}{|A|}\left(f'\left(m^2U_{cc} - 2mU_{cW} + U_{WW}\right)\right) \tag{4.23}$$

where

$$|A| = \begin{vmatrix} m^2U_{cc} - 2mU_{cW} + U_{WW} & mU_{cc} - U_{cW} \\ mU_{cc} - U_{cW} & U_{cc} + \varphi f'' \end{vmatrix} \tag{4.24}$$

Since, $sgn\,|A| > 0$, then $sgn\frac{\partial I}{\partial \phi} < 0$, from Eq. (4.23). However, $sgn\frac{\partial I}{\partial x}$ is not definite. If U_{cW} is nearly equal to 0, ($U_{cW} = \frac{dU_c}{dW} \fallingdotseq 0$); in other words, if the change in water rates does not affect the marginal utility of consumption, then as per Eq. (4.24), $sgn\frac{\partial I}{\partial x} > 0$.

Next, we consider the stationary state, that is, the case wherein $\dot{\phi} = 0$ and $\dot{x} = 0$ hold. Therefore, in the former case, Eq. (4.15) means

$$\rho + \delta - nf'\left(I\left(x, \varphi\right)\right) = 0 \tag{4.25}$$

In the latter, from Eq. (4.3), we obtain

$$f\left(I\left(x, \varphi\right)\right) - \delta x = 0 \tag{4.26}$$

Totally differentiating Eqs. (4.25) and (4.26),

$$\frac{\partial I}{\partial x}dx + \frac{\partial I}{\partial \varphi}d\varphi = 0$$

$$\left(f'\frac{\partial I}{\partial x} - \delta\right)dx + f'\frac{\partial I}{\partial \varphi}d\varphi = 0$$

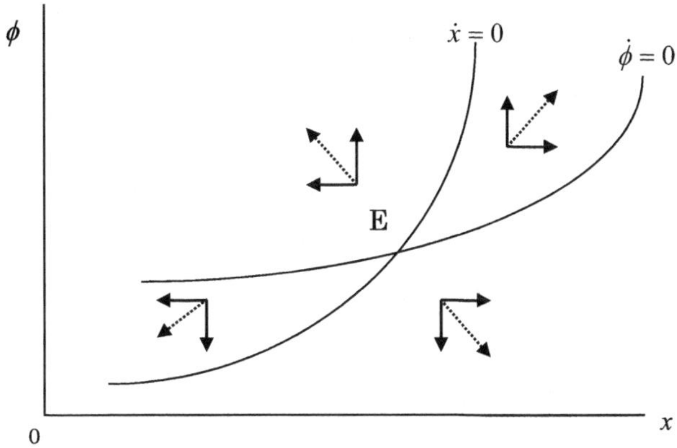

Fig. 4.3 Case 1

Then,

$$\frac{d\varphi}{dx}\bigg|_{\dot{\varphi}=0} = -\frac{\partial I/\partial x}{\partial I/\partial \varphi} \tag{4.27}$$

$$\frac{d\varphi}{dx}\bigg|_{\dot{x}=0} = -\frac{f'\partial I/\partial x - \delta}{f'\partial I/\partial \varphi} \tag{4.28}$$

By $sgn\frac{\partial I}{\partial \phi} < 0$ and $sgn\frac{\partial I}{\partial x} > 0$, we obtain $sgn\frac{d\phi}{dx}\Big|_{\dot{\phi}=0} > 0$ from Eq. (4.27). However, $sgn\frac{d\phi}{dx}|_{\dot{x}=0}$ is not definite. Accordingly, the cases can be classified into the following three cases.

	$\dot{\phi} = 0$	$\dot{x} = 0$		
case1	Rising	Rising	The slope($\dot{\phi} = 0$)<the slope($\dot{x} = 0$)	Fig. 4.3
case2	Rising	Rising	The slope($\dot{\phi} = 0$)>the slope($\dot{x} = 0$)	Fig. 4.4
case3	Rising	Declining		Fig. 4.5

From the figures above, we find that steady-state solutions exist in Case 2 and Case 3, and they are saddle points.

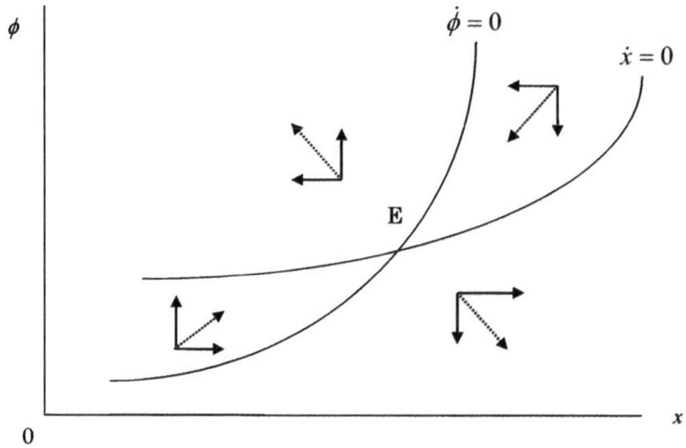

Fig. 4.4 Case 2

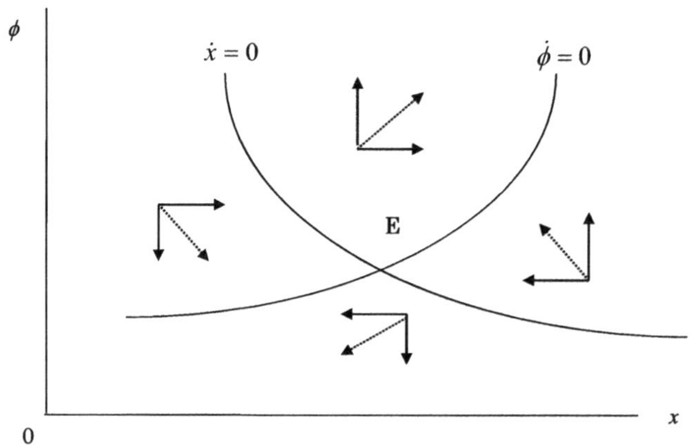

Fig. 4.5 Case 3

4.5 Lump-Sum Tax

In the previous section, we did not consider individual behavior to consider social optimization. However, in this section we assume that consumption and use of tap water are determined by individual optimization behavior. In addition to the water fee, the municipal authority imposes a lump-sum tax T_t which is expended for forest recharge every period. As for the water fee, both the basic fixed fee and the metered fee are collected. We assume that the metered rate per cubic meter is m, the basic flat rate fee in period t is $K - nx_t$, and the metered fee is mW_t.

In this case, the maximizing problem of family in period t is

$$\max_{c_t, W_t} U\,(c_t,\, W_t)$$

subject to

$$Y = c_t + mW_t + K - nx_t + T_t \tag{4.29}$$

The first-order conditions are

$$U_c = \lambda \tag{4.30}$$

$$U_c = \lambda \tag{4.31}$$

$$Y = c_t + mW_t + K - nx_t + T_t \tag{4.32}$$

where λ is a Lagrange multiplier.

We obtain an individual's demand functions of consumption and water with respect to the rate of water and lump-sum tax rate by solving the optimal conditions in Eqs. (4.30), (4.31), and (4.32) as follows:

$$c_t = c\,(m, Y - K + nx_t - T_t) \tag{4.33}$$

$$W_t = c\,(m, Y - K + nx_t - T_t) \tag{4.34}$$

The municipality determines the level of lump-sum taxes in each period to maximize the social welfare function, subjected to the demand functions of citizens in each period. Then, our maximization problem is formulated as

$$\max_T \int_0^\infty U\left(c\,(m, Y - K + nx_t)\,, W\,(m, Y - K + nx_t)\right) e^{-\rho t} dt$$

subject to

$$\dot{x}_t = f\,(T_t) - \delta x_t$$

$$x_0 = \overline{x}$$

Let the present value Hamiltonian H' be

$$H' = U\left(c\,(m, Y - K + nx_t - T_t)\,, W\,(m, Y - K + nx_t - T_t)\right) + \xi \\ \{f\,(T_t) - \delta x_t\}\,, \tag{4.35}$$

where ξ is an auxiliary variable.

The optimal solutions in this problem must satisfy the following conditions:

$$-U_c\left(\frac{\partial c}{\partial Y}\right)-U_W\left(\frac{\partial W}{\partial Y}\right)+\xi f'=0 \tag{4.36}$$

$$\dot{\xi}=\rho\xi+U_c\left(\frac{\partial c}{\partial Y}n\right)+U_W\left(\frac{\partial W}{\partial Y}n\right)-\delta\xi \tag{4.37}$$

Equations (4.36) and (4.37) are derived to meet the following equation by equilibrium conditions of consumers:

$$\lambda+\xi f'=0 \tag{4.38}$$

$$\dot{\xi}=\rho\xi+n\lambda+\delta\xi \tag{4.39}$$

From Eq. (4.38), we obtain

$$\lambda=-\xi f' \tag{4.40}$$

Substituting this into Eq. (4.39) and arranging, we get

$$\dot{\xi}=\xi\left(\rho+\delta-nf'\right) \tag{4.41}$$

Equation (4.41) coincides with the optimal condition in Eq. (4.16).

Therefore, we conclude that even if consumers independently decide the use of tap water, the method of forest conservation by lump-sum tax can lead to an optimal solution.

4.6 The Forest Conservation Policy of Toyota City in Aichi Prefecture

The Toyota City system levies 1 yen per m^3 (1.08 yen including tax) as a forest recharge fund by adding it to the water fee. We examine in this section whether this scheme can attain the social optimal in the same supposition as the former section.

We assume that the marginal cost of water supply is m and tax surcharge is equal to s yen on water per 1 m^3. Residents then pay $\{(m+s)W_t+K-nx_t\}$ for water supply W_t in period t. The fund of the forest conservation increases by sW_t in period t.

Individual families in this city decide consumptions and water demands to maximize their utility by taking the rate and surcharge on water as given. The maximizing problem of each family is formulated as

$$\max_{c_t, W_t} U\left(c_t, W_t\right)$$

subject to

$$Y = c_t + (m + s)\, W_t + K - nx_t$$

The first-order conditions are

$$U_c = \eta \tag{4.42}$$

$$U_W = (m + s)\, \eta \tag{4.43}$$

$$Y = c_t + (m + s)\, W_t + K - nx_t \tag{4.44}$$

From Eqs. (4.42), (4.43) and (4.44), we obtain

$$c_t = c\left(m + s, Y - K + nx_t\right) \tag{4.45}$$

$$W_t = W\left(m + s, Y - K + nx_t\right) \tag{4.46}$$

Equations (4.45) and (4.46) are the demand functions for consumption and water, respectively.

The city decides the rate and surcharge on water to maximize social welfare with the demand function for consumption and water as a given. The maximizing problem for the government in this case is

$$\max_{s} \int_0^{\infty} U\left(c\left(m + s, Y - K + nx_t\right), W\left(m + s, Y - K + nx_t\right)\right) e^{-\rho t} dt$$

subject to

$$\dot{x}_t = f\left(sW_t\right) - \delta x_t$$

$$x_0 = \bar{x}.$$

The present value of the Hamiltonian H'' is defined as

$$H'' = U\left(c\left(m+s, Y-K+nx_t\right), W\left(m+s, Y-K+nx_t\right)\right) + \psi \\ \left\{f\left(sW_t\right) - \delta x_t\right\}, \tag{4.47}$$

where ψ is an auxiliary variable.

The necessary conditions of this problem are

$$U_c\frac{\partial c}{\partial s} + U_W\frac{\partial W}{\partial s} + \psi f'\left(W_t + \frac{\partial W}{\partial s}s\right) = 0 \tag{4.48}$$

$$\dot{\psi} = \rho\psi - U_c\frac{\partial c}{\partial Y}n - U_W\frac{\partial W}{\partial Y}n + \psi\delta \tag{4.49}$$

Equations (4.48) and (4.49) are reduced from the equilibrium conditions of individuals, and Eqs. (4.10), (4.11), and (4.12) as follows:

$$\eta\frac{\partial c}{\partial s} + \eta\left(m+s\right)\frac{\partial W}{\partial s} + \psi f'\left(W_t + \frac{\partial W}{\partial s}s\right) = 0 \tag{4.50}$$

$$\dot{\psi} = \rho\psi - \eta\frac{\partial c}{\partial Y}n - \eta\left(m+s\right)\frac{\partial W}{\partial Y}n + \psi\delta \tag{4.51}$$

Equation (4.50) is transformed into Eq. (4.52)

$$-\eta W_t + \psi f'\left(W_t + \frac{\partial W}{\partial s}s\right) = 0, \tag{4.52}$$

and Eq. (4.51) is transformed into Eq. (4.53)

$$\dot{\psi} = \rho\psi - \eta n\left\{\frac{\partial c}{\partial Y} + \left(m+s\right)\frac{\partial W}{\partial Y}\right\} + \psi\delta\dot{\psi} = \left(\rho+\delta\right)\psi - \eta n \tag{4.53}$$

Further, Eq. (4.52) is transformed into Eq. (4.54):

$$\eta = \psi f' (1 + \varepsilon_t) \tag{4.54}$$

Here, we define $\epsilon_t = \frac{s}{W_t}\frac{\partial W}{\partial s} = \frac{s}{W_t}\frac{\partial W}{\partial (m+s)}$, and then ϵ_t is the price elasticity of demand for water in period t. By substituting Eq. (4.54) into Eq. (4.53), and rearranging, we get

$$\dot{\psi} = \psi \left\{ \rho + \delta - nf' (1 + \varepsilon_t) \right\} \tag{4.55}$$

Equation (4.55) coincides with the optimal condition in Eq. (4.16) if the price elasticity of the tap water demand is zero. Generally, the price elasticity of water supply is extremely small but not zero. Hence, so the Toyota scheme assumed in this section does not achieve the optimum.

Restricting to the steady state, Eq. (4.55) is reduced to

$$nf' (1 + \varepsilon_t) = \rho + \delta \tag{4.56}$$

or

$$nf' = \frac{\rho + \delta}{1 + \varepsilon_t} \tag{4.57}$$

As depicted in Fig. 4.6, the forest recharge fund level that satisfies Eq. (4.56) is lower than the level that satisfies Eq. (4.16). Therefore, the long-term forest recharge level is shown to be too low in the Toyota City system. Although the degree of distortion depends on the price elasticity of water demand, the price elasticity of water demand is extremely low in general. A similar trend was seen in OECD (1999), and neither the fee nor the income had a small influence on water consumption.

From the above, we conclude that the funding scheme for forest conservation using lump-sum tax is most desirable and the Toyota scheme can attain the almost best solution.

In addition, the cost of forest recharge does not need to depend on a specific financial source. For instance, Nakayama et al. (2017) examined the efficiency of a policy that covers the cost with a lump-sum tax, income tax, or consumption tax imposed on pollution sources.

4.7 Conclusion

In this paper, we analyzed the lump-sum-type forest environment tax and the water source fund of Toyota City. When considering efficiency in resource allocation, we can conclude that the lump-sum tax method is superior to the Toyota City method.

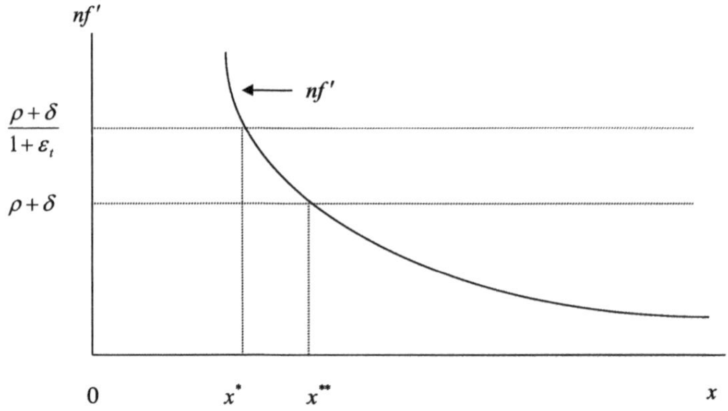

x^* : the steady state of forest stock of the Toyota scheme
x^{**} : the steady state of forest stock

Fig. 4.6 The steady state of forest stock

However, when considering impartiality and beneficiary burden, the Toyota City system, which depends on usage amount, is superior to the lump-sum tax system, wherein a fixed amount is collected regardless of the amount of benefits.

Tax revenues for the forest environment tax newly introduced in Japan are estimated to reach 62 billion yen. However, the new tax has received several criticisms. Urban residents who bear heavy tax are less likely to receive benefits of forest. Further, the difference between the forest environment tax by central government and the forest environment taxes already introduced by local governments is not clear. These taxes imposed by both central and local governments are afraid of double taxation. In addition, the forest environment tax has characteristics like subsidies only for specific industries, such as forestry. Some critics have also argued that if it is a subsidy, it is not appropriate to allocate it from the budget without collecting it as national tax.

To obtain a broader understanding of the residents, we must clarify the relationship between benefits and burdens. Other potential problems should also be addressed in future, such as the ways in which the cost of the forest environmental tax should be borne, the collection method and its uses, the evaluation of implementation projects with tax on tax revenues, and the basin governance that arises from the watershed in surrounding forests spreading to multiple areas.

Acknowledgments The authors highly appreciate financial supports from the Japan Society for the Promotion of Science (Grant-in-Aid for Scientific Research (C), 18 K01631) and Chukyo University Institute of Economics (P15002).

This paper is an extended version of Nakayama et al. (2016) and Nakayama and Shirai (2009). We are grateful to the Toyota City Waterworks Bureau that provided the materials for this study.

References

Amacher GS, Brazee RJ (2013) Designing forest taxes with varying government preferences and budget targets. J Environ Econ Manag 32(3):323–340. Elsevier. https://doi.org/10.1006/jeem.1997.0970

Choo SH, Lee J, Roberts RK, English BC (2017) Evaluating a tax-based subsidy approach for forest carbon sequestration. Env Conserv, Int J Interdisciplinary Env Sci 44(3):234–243. Cambridge University Press. https://doi.org/10.1017/S0376892917000078

Cushing TL, Newman D (2018) Analysis of relative tax burden on nonindustrial private forest landowners in the Southeastern United States. J For 116(3):4. 228–235, 176–182, Elsevier. https://doi.org/10.1093/jofore/fvx013

Gong P, Löfgren KG (2013) Forest taxation, encyclopedia of energy, natural resource, and environmental economics, vol 2. Elsevier, Amsterdam, pp 176–182. https://doi.org/10.1016/B978-0-12-375067-9.00050-4

Hall RC (2017) The forest-tax problem and its solution summarized (classic report). Forgotten Press

Harada H (2001) Preservation of upstream forest with one ton 1 yen Toyota City water supply conservation fund and project initiatives. Monthly Autonomous Research August 2001 Issue: 56–62 (in Japanese)

Honda H (2007) The problem of "forest fund" as eco-goods schemes –based on the point of view of "the nonprofit making business model of social entrepreneur". Stud Reg Sci 18:51–62

Kamiya K (2000) Case study: about Toyota City water supply conservation fund, Monthly Public Enterprise September 2000 Issue 32(6):72–77 (in Japanese)

Kimbell AR, Hickman K, Brown H (2010) How taxes affect America's private forestland owners? J For 108(2):1, 93–97. https://doi.org/10.1093/jof/108.2.93

Liu J, Wu F (2017) Forest carbon sequestration subsidy and carbon tax as part of China's forestry policies. Forests 8(3):58. https://doi.org/10.3390/f8030058

Nakayama K, Shirai M (2009) Burden of public investment costs for forest conservancy in preservation of water supply sources –analysis of the effects of environmental taxes on the destruction of forests. Stud Reg Sci 39(3):553–565. (in Japanese)

Nakayama K, Shirai M, Yamada M (2007) Economic evaluation of Toyota City water supply conservation project in Aichi prefecture II -analysis of forest environmental tax by optimization problem. Chukyo Econ Rev 18:65–77. (in Japanese)

Nakayama K, Shirai M, Matsumoto A (2016) Environmental preservation policy. Chukyo Univ Inst Econ, Discuss Paper Ser 1609:1–12

Nakayama K, Shirai M, Matsumoto A (2017) Environment conservation policy focusing on forest recharge, in new development of economic theory, application, empirical analysis, the Institute of Economic Research, Chuo University, Study Series 72, pp 145–162 (in Japanese)

OECD (1999) The price of water –trends in OED countries-, OECD Library. https://doi.org/10.1787/9789264173996-en

Siegel WC (2010) Comments on "how do taxes affect America's private forestland owners?" J For 108(2): 1, 98–99, Oxford Academic Press. https://doi.org/10.1093/jof/108.2.98

Stavins RN, Richards KR (2005) The cost of U.S. forest-based carbon sequestration. Prepared by the pew center on global climate change. http://www.c2es.org/publications/cost-us-forest-based-carbon-sequestration

Part II
Environment: Empirical Approach

Chapter 5
An Interactive Rural–Urban–Natural Environment Model of a City with Illegal Settlements in a Flood-Prone Area

Yuzuru Miyata and Hiroyuki Shibusawa

Abstract This paper presents a theoretical framework for rural and urban interactive models regarding the existence of illegal settlements in flood-prone areas of a city. Considering the natural environment and the externality of the forest, there are two types of household in the city: high and low income. Household and firm land use in urban and rural areas, including flood-prone areas, is analyzed, and policy implications are provided.

Keywords Rural and urban economic model · Natural environment · Flood-prone areas

5.1 Introduction

As the world continues to urbanize, cities offer attractive spaces for their inhabitants. They also contribute to the efficiency of productive activity and infrastructure investment. However, there is an extreme concentration of populations and economic activities in ever increasingly smaller spaces. This results in a range of social difficulties, such as housing shortages, traffic congestion, poverty, slum dwelling, and high crime and magnifies the impact of natural disasters.

Indonesia is a country that consists of approximately 17,000 islands. The urban population of Jakarta surpasses 10 million people. In many major cities, traffic and infrastructure maintenance requirements cannot meet population demands, creating serious urban problems.

Y. Miyata (deceased)
Toyohashi, Aichi, Japan

H. Shibusawa (✉)
Department of Architecture and Civil Engineering, Toyohashi University of Technology, Toyohashi, Aichi, Japan
e-mail: hiro-shibu@tut.jp

K. Nakayama, Y. Miyata (eds.), *Theoretical and Empirical Analysis in Environmental Economics*, New Frontiers in Regional Science: Asian Perspectives 34, https://doi.org/10.1007/978-981-13-2363-8_5

Table 5.1 Population and economic indicators in Indonesia, 2014

Population	$249,865,631 Million
Urban	52.25%
Rural	47.75%
Urban population growth	2.68%
Population density	137.9%
GDP	$868,345.645 Million
GDP per capita	$3475.25

Source: Prevention Web (www.preventionweb.net)

Many cities in Indonesia undergo rapid population growth. As shown in Table 5.1, Indonesia experienced a 2.68% increase in its 2014 urban population. The percentage of the population living in urban areas is now 52.25%, with evidence showing significant movement from rural to urban areas. Additionally, the recent growth has resulted in a separation of wealth. Approximately 200 million people (80% of the population) are in the lower-income class. Furthermore, natural disasters (e.g., earthquakes, volcanos, floods, landslides, tsunami, thinning forests, and forest fires) are exacerbated by Indonesia's geographical location. Monsoons cause very wet rainy seasons, many floods, and resultant inundation damage each year.

Urban and rural interaction is an important issue at both local and global levels. Agricultural development has accelerated the growth of many cities: however, this growth has a pervasive impact on agricultural production within the global economy (Leeuwen and Nijkamp 2006). The links between rural and urban areas are significantly influenced by the global economy. During the economic crisis of 1997 and the implementation of a free-market policy, urban and rural linkages changed. The crisis had a detrimental impact on industrial sectors in urban areas but was advantageous to agricultural sectors in rural areas (Miyata et al. 2018). Urban economics is a relatively young field, developed by many economics, engineering, and geography scholars since the 1960s (Alonso, 1964; Muth 1969; Mills 1967; Solow 1973; Richardson 1977; Fujita, 1989; Anas et al. 1998; Black and Henderson 1999; Fujita et al. 1999; Higano and Shibusawa 1999; Shibusawa 2000; Lucas and Rossi–Hansberg 2002; and Puu 2003). Permana and Miyata (2009) showed a 2-dimensional (2D) urban economic model by introducing bid-rent functions for both firms and households and then applying the theory of partial differential equations to those (Courant and Hilbert 1953, 1962), including the asymptotic expansion method (Hörmander 1990a, b). Permana and Miyata (2012a) showed a partial equilibrium urban economics model, explaining the existence of illegal settlements in flood-prone areas in Palangkaraya City in the central Kalimantan province and introducing an expected damage rate on household assets. Applying this model, one could derive scenarios where the bid-rents of low-income households (L.I.H.) grow larger than those of high-income households (H.I.H.) in flood-prone areas. This pattern contradicts traditional urban economics.

Permana and Miyata (2012b) went further to extend the partial equilibrium model into a general equilibrium model. They then developed a 2D city model applying Miyata's achievements (Permana and Miyata 2009; Miyata 2011). However, the study region, Palangkaraya City and its surrounding area, showed a complicated interaction of natural environment and human activities. Therefore, this article aims to describe a rural and urban economic model that better considers the natural environment, in an attempt to surpass prior literature.

We describe the assumptions of our model in Sect. 5.2. Household and firm behaviors are formulated in Sects. 5.3, 5.4, 5.5, and 5.6. Section 5.7 describes the Hamiltonian Property. Market and balance equations are defined in Sects 5.8, 5.9, and 5.10. Behaviors of the local government and absentee landowners are described in Sect. 5.11. In Sects. 5.12 and 5.13, the behaviors of the transport agent for commodity and labor are explained. Commodity prices and wage rates in a plane space are defined in Sect. 5.14. Equilibrium conditions are presented in Sect. 5.15. A comparative static analysis is conducted in Sect. 5.16. Finally, Sect. 5.17 concludes this paper.

5.2 Assumptions of the Model

Our model is based on the following assumptions:

1. The study region consists of an urban area (i.e., Palangkaraya City) and its surrounding rural areas. The city shape is assumed to be a disk where there are three flood-prone areas. Non-flood-prone areas are called "normal land." This normal land is assumed to have no flood risk, while the flood-prone areas face flood risks with known probabilities. The rural areas are specified as dimensionless (i.e., they have no spatial structures).
2. There are two types of household (i.e., H.I.H.s and L.I.H.s) in the city. H.I.H.s are assumed to exist on normal land, whereas L.I.H.s exist in the flood-prone areas. The city is assumed to be closed for H.I.H.s and open for L.I.H.s. This reflects the utility of a H.I.H. in urban areas being much higher than that of rural areas. Thus, there is no incentive to live outside the city. L.I.H.s in rural areas expect better services and utilities in the city, and as such they want to live in the city. However, most people remain in the flood-prone areas because of the vast income gap. Thus, only L.I.H.s exist in the rural areas. The number of H.I.H.s is denoted by N_1; that of L.I.H.s in the city is denoted by N_2; and that of L.I.H.s in rural areas is denoted by N_3.
3. We consider different types of commercial entities in the study region. In the city, all firms are assumed to be homogeneous, producing a single type of good (i.e., urban goods). In the rural area, agricultural, forest, and rural goods are produced. The number of firms in the city is M, whereas firms in the rural area are aggregated into the three types.
4. Land in the city is owned by absentee landowners who reside outside the city. The supply of each type of land is exogenously given. There is a unique local government in the city that rents all land owned by absentee landowners to households and firms at market cost. Thus, wealth is redistributed from the poor to the rich.

5. Capital stock in the city is assumed to be freely mobile across firms. Thus, the capital rate of return is uniquely determined, irrespective of the firms' locations. The capital service is assumed to be *numerare*.
6. The parameters in the locational potential function for each firm are sufficiently large. In this case, a simple von Thünen ring becomes an equilibrium urban configuration (von Thünen 1826; Miyata 2011).

5.3 Firms' Behavior in the City

The production function of a firm at location $\boldsymbol{x} = (x_1, x_2)$ in the city is specified as a Cobb–Douglas–CES type with a homogeneous degree of unity. Additionally, an agglomeration economy is considered. The agglomeration economy is represented by a locational potential function, $\Omega(\boldsymbol{x})$, introduced by Fujita and Ogawa (1982). It is defined as follows:

$$\Omega(\boldsymbol{x}) = \iint_A \mu b(\boldsymbol{y}) \exp(-\omega \|\boldsymbol{x} - \boldsymbol{y}\|) \, dy_1 dy_2, \tag{5.1}$$

where:

A: city area
$b(\boldsymbol{y})$: density of firms at location $\boldsymbol{y}$
μ: monetary conversion parameter in locational potential
ω: parameter expressing the effects of distance between different points
$\|\boldsymbol{x} - \boldsymbol{y}\|$: distance between locations $\boldsymbol{x}$ and $\boldsymbol{y}$

The production of a firm at location $\boldsymbol{x}$ may be written as follows:

$$\begin{aligned} q_1(\boldsymbol{x}) = {} & \Omega(\boldsymbol{x}) q_{A1} \prod_{i=1}^{4} q_{i1}(\boldsymbol{x})^{\alpha_{i1}} \\ & \times \left(\left[\zeta^{\frac{1}{\sigma_1}} ld_1(\boldsymbol{x})^{\frac{\sigma_1 - 1}{\sigma_1}} + (1-\zeta)^{\frac{1}{\sigma_1}} ld_2(\boldsymbol{x})^{\frac{\sigma_1 - 1}{\sigma_1}} \right]^{\frac{\sigma_1}{\sigma_1 - 1}} \right)^{\alpha_{l1}} \\ & \times kd_1(\boldsymbol{x})^{\alpha_{k1}} m_{B1}(\boldsymbol{x})^{\alpha_{m1}} (u_{U0})^{\alpha_{U01}} (n_{R3})^{\alpha_{R31}}, \end{aligned} \tag{5.2}$$

where:

$q_1(\boldsymbol{x})$: output of a firm at location $\boldsymbol{x}$
q_{A1}: efficient parameter
$q_{11}(\boldsymbol{x})$: intermediate input of urban goods at location $\boldsymbol{x}$
$q_{21}(\boldsymbol{x})$: intermediate input of agricultural products at location $\boldsymbol{x}$
$q_{31}(\boldsymbol{x})$: intermediate input of forestry products at location $\boldsymbol{x}$
$q_{41}(\boldsymbol{x})$: intermediate input of rural goods at location $\boldsymbol{x}$
$ld_1(\boldsymbol{x})$: input of labor of high-income type at location $\boldsymbol{x}$
$ld_2(\boldsymbol{x})$: input of labor of low-income type at location $\boldsymbol{x}$

$kd_1(\boldsymbol{x})$: capital input at location $\boldsymbol{x}$
$mB_1(\boldsymbol{x})$: land input at location $\boldsymbol{x}$
u_{U0}: natural environmental level in the city (e.g. air and water)
n_{R3}: forest volume in the rural area
ζ: share parameter
σ_1: elasticity of substitution $(0 < \sigma_1 < 1)$
$\alpha_{i1}, \alpha_{l1}, \alpha_{k1}, \alpha_{m1}, \alpha_{U01}, \alpha_{R31}$: elasticity parameters $(\alpha_{11} + \alpha_{21} + \alpha_{31} + \alpha_{41} + \alpha_{l1} + \alpha_{k1} + \alpha_{m1} = 1)$.

We assume that each firm is a price taker for commodities and production factors. The impact of production activity by each firm on the forest and wider natural environment is negligible. The firms' locational equilibrium condition is one in which the profit of each firm is equalized at every point. Owing to the linear homogeneity of degree in each firm's technology, the equilibrium profit in each firm becomes zero. Then, via the bid-rent function, conditional demands for intermediate goods, the two types of labor, capital stock, and bid-max lot size are obtained.

$$g_{B1}(\boldsymbol{x}) = \max \left[\frac{1}{m_{B1}(\boldsymbol{x})} \left\{ p_1(\boldsymbol{x}) q_1(\boldsymbol{x}) - \sum_{i=1}^{4} p_i(\boldsymbol{x}) q_{i1}(\boldsymbol{x}) - \sum_{i=1}^{2} w_i(\boldsymbol{x}) ld_i(\boldsymbol{x}) - r_1(\boldsymbol{x}) kd_1(\boldsymbol{x}) \right\} \right], \tag{5.3}$$

with respect to $q_{i1}(\boldsymbol{x})$, $ld_i(\boldsymbol{x}) kd_1(\boldsymbol{x})$, and $m_{B1}(\boldsymbol{x})$,
subject to:

$$q_1(\boldsymbol{x}) = \Omega(\boldsymbol{x}) q_{A1} \prod_{i=1}^{4} q_{i1}(\boldsymbol{x})^{\alpha_{i1}} \times \left(\left[\zeta^{\frac{1}{\sigma_1}} ld_1(\boldsymbol{x})^{\frac{\sigma_1 - 1}{\sigma_1}} + (1-\zeta)^{\frac{1}{\sigma_1}} ld_2(\boldsymbol{x})^{\frac{\sigma_1 - 1}{\sigma_1}} \right]^{\frac{\sigma_1}{\sigma_1 - 1}} \right)^{\alpha_{l1}} \times kd_1(\boldsymbol{x})^{\alpha_{k1}} m_{B1}(\boldsymbol{x})^{\alpha_{m1}} (u_{U0})^{\alpha_{U01}} (n_{R3})^{\alpha_{R31}}, \tag{5.4}$$

$$\pi_{B1}(\boldsymbol{x}) = 0, \tag{5.5}$$

where:

$p_1(\boldsymbol{x})$: price of urban good at location $\boldsymbol{x}$
$p_2(\boldsymbol{x})$: price of agricultural product at location $\boldsymbol{x}$
$p_3(\boldsymbol{x})$: price of forestry product at location $\boldsymbol{x}$
$p_4(\boldsymbol{x})$: price of rural good at location $\boldsymbol{x}$
$w_1(\boldsymbol{x})$: wage rate of a H.I.H. at location $\boldsymbol{x}$
$w_2(\boldsymbol{x})$: wage rate of a L.I.H. at location $\boldsymbol{x}$
$r_1(\boldsymbol{x})$: capital return rate at location $\boldsymbol{x}$

$g_{B1}(\boldsymbol{x})$: bid-rent by a firm at location $\boldsymbol{x}$
$\pi_{B1}(\boldsymbol{x})$: profit in a firm at location $\boldsymbol{x}$

To obtain the bid-rent function and the bid-max lot size of a firm, we use the cost function.

$$C\left(q_1\left(\boldsymbol{x}\right)\right) \equiv \min\left[\sum_{i=1}^{4} p_i\left(\boldsymbol{x}\right) q_{i1}\left(\boldsymbol{x}\right) + \sum_{i=1}^{2} w_i\left(\boldsymbol{x}\right) ld_i\left(\boldsymbol{x}\right) + r_1\left(\boldsymbol{x}\right) kd_1\left(\boldsymbol{x}\right) \right. \\ \left. + g_{B1}\left(\boldsymbol{x}\right) m_{B1}\left(\boldsymbol{x}\right)\right], \tag{5.6}$$

with respect to $q_{i1}(\boldsymbol{x})$, $ld_i(\boldsymbol{x})kd_1(\boldsymbol{x})$, and $m_{B1}(\boldsymbol{x})$,
subject to:

$$q_1\left(\boldsymbol{x}\right) = \Omega\left(\boldsymbol{x}\right) q_{A1} \prod_{i=1}^{4} q_{i1}(\boldsymbol{x})^{\alpha_{i1}} \\ \times \left(\left[\zeta^{\frac{1}{\sigma_1}} ld_1(\boldsymbol{x})^{\frac{\sigma_1-1}{\sigma_1}} + (1-\zeta)^{\frac{1}{\sigma_1}} ld_2(\boldsymbol{x})^{\frac{\sigma_1-1}{\sigma_1}}\right]^{\frac{\sigma_1}{\sigma_1-1}}\right)^{\alpha_{l1}} \\ \times kd_1(\boldsymbol{x})^{\alpha_{k1}} m_{B1}(\boldsymbol{x})^{\alpha_{m1}} (u_{U0})^{\alpha_{U01}} (n_{R3})^{\alpha_{R31}}. \tag{5.7}$$

Then the cost function is solved as follows:

$$C\left(q_1\left(\boldsymbol{x}\right)\right) = \frac{q_1(\boldsymbol{x})}{\Omega(\boldsymbol{x}) q_{A1} (u_{U0})^{\alpha_{U01}} (n_{R3})^{\alpha_{R31}}} \\ \times \prod_{i=1}^{4} \left[\frac{p_i(\boldsymbol{x})}{\alpha_{i1}}\right]^{\alpha_{i1}} \\ \times \left[\frac{\left[\zeta w_1(\boldsymbol{x})^{1-\sigma_1} + (1-\zeta) w_2(\boldsymbol{x})^{1-\sigma_1}\right]^{\frac{1}{1-\sigma_1}}}{\alpha_{l1}}\right]^{\alpha_i} \\ \times \left[\frac{r_1(\boldsymbol{x})}{\alpha_{k1}}\right]^{\alpha_{k1}} \left[\frac{g_{B1}(\boldsymbol{x})}{\alpha_{m1}}\right]^{\alpha_{m1}}. \tag{5.8}$$

Because the equilibrium profit is zero, the following equation holds:

$$p_1\left(\boldsymbol{x}\right) q_1\left(\boldsymbol{x}\right) = C\left(q_1\left(\boldsymbol{x}\right)\right). \tag{5.9}$$

Then, the bid-rent function of the firm is solved:

$$g_{B1}\left(\boldsymbol{x}\right) = \alpha_{m1}\left[p_1\left(\boldsymbol{x}\right) \Omega\left(\boldsymbol{x}\right) q_{A1} (u_{U0})^{\alpha_{U01}} (n_{R3})^{\alpha_{R31}}\right]^{\frac{1}{\alpha_{m1}}} \\ \times \prod_{i=1}^{4} \left[\frac{\alpha_{i1}}{p_i(\boldsymbol{x})}\right]^{\frac{\alpha_{i1}}{\alpha_{m1}}} \\ \times \left[\frac{\alpha_{l1}}{\left[\zeta w_1(\boldsymbol{x})^{1-\sigma_1} + (1-\zeta) w_2(\boldsymbol{x})^{1-\sigma_1}\right]^{\frac{1}{1-\sigma_1}}}\right]^{\frac{\alpha_{i1}}{\alpha_{m1}}} \\ \times \left[\frac{\alpha_{k1}}{r_1(\boldsymbol{x})}\right]^{\frac{\alpha_{k1}}{\alpha_{m1}}}, \tag{5.10}$$

$$q_{i1}(\boldsymbol{x}) = \frac{\alpha_{i1} p_1(\boldsymbol{x}) q_1(\boldsymbol{x})}{p_i(\boldsymbol{x})} \quad (i = 1, 2, 3, 4), \tag{5.11}$$

$$ld_1(\boldsymbol{x}) = \frac{\alpha_{l1} \zeta p_1(\boldsymbol{x}) q_1(\boldsymbol{x})}{w_1(\boldsymbol{x})^{\sigma_1} \left[\zeta w_1(\boldsymbol{x})^{1-\sigma_1} + (1-\zeta) w_2(\boldsymbol{x})^{1-\sigma_1}\right]^{\frac{\sigma_1}{\sigma_1 - 1}}}, \tag{5.12}$$

$$ld_2(\boldsymbol{x}) = \frac{\alpha_{l1} (1-\zeta) p_1(\boldsymbol{x}) q_1(\boldsymbol{x})}{w_2(\boldsymbol{x})^{\sigma_1} \left[\zeta w_1(\boldsymbol{x})^{1-\sigma_1} + (1-\zeta) w_2(\boldsymbol{x})^{1-\sigma_1}\right]^{\frac{\sigma_1}{\sigma_1 - 1}}}, \tag{5.13}$$

$$kd_1(\boldsymbol{x}) = \frac{\alpha_{k1} p_1(\boldsymbol{x}) q_1(\boldsymbol{x})}{r_1(\boldsymbol{x})}, \tag{5.14}$$

$$\begin{aligned} m_{B1}(\boldsymbol{x}) &= \left[\Omega(\boldsymbol{x}) q_{A1} (u_{U0})^{\alpha_{U01}} (n_{R3})^{\alpha_{R31}}\right]^{\frac{-1}{\alpha_{m1}}} \prod_{i=1}^{4} \left[\frac{\alpha_{i1}}{p_i(\boldsymbol{x})}\right]^{\frac{\alpha_{i1}}{\alpha_{m1}}} \\ &\times \left[\frac{\alpha_{l1}}{\left[\zeta w_1(\boldsymbol{x})^{1-\sigma_1} + (1-\zeta) w_2(\boldsymbol{x})^{1-\sigma_1}\right]^{\frac{1}{1-\sigma_1}}}\right]^{\frac{\alpha_{i1}}{\alpha_{m1}}} \\ &\times \left[\frac{\alpha_{k1}}{r_1(\boldsymbol{x})}\right]^{\frac{\alpha_{k1}}{\alpha_{m1}}} q_1(\boldsymbol{x}). \end{aligned} \tag{5.15}$$

5.4 Households' Behavior in the City

The household utility functions in both types of households at location $\boldsymbol{x}$ are expressed as follows:

$$\begin{aligned} &u_i\left(c_{1i}(\boldsymbol{x}), c_{2i}(\boldsymbol{x}), c_{3i}(\boldsymbol{x}), c_{4i}(\boldsymbol{x}), c_{Fi}(\boldsymbol{x}), m_{Hi}(\boldsymbol{x}); n_{U0}, n_{R3}\right) \\ &\equiv \frac{\prod_{k=1}^{4} c_{ki}(\boldsymbol{x})^{\beta_{k1}} c_{Fi}(\boldsymbol{x})^{\beta_{F1}} m_{Hi}(\boldsymbol{x})^{\beta_{m1}} (n_{U0})^{\beta_{U01}} (n_{R3})^{\beta_{R31}}}{1 + c(n_{R3})^{-\omega} Y_i(\boldsymbol{x})^{\varepsilon}} \quad (i = 1, 2) \end{aligned} \tag{5.16}$$

$$Y_i(\boldsymbol{x}) \equiv w_i(\boldsymbol{x}) + r_1(\boldsymbol{x}) ks_i(\boldsymbol{x}) + \pi_{Hi}(\boldsymbol{x}), \tag{5.17}$$

where:

$i : i = 1$ for a H.I.H. and $i = 1$ for a L.I.H.
u_i: household utility function at location $\boldsymbol{x}$

$c_{1i}(\boldsymbol{x})$: household consumption of urban goods at location $\boldsymbol{x}$
$c_{2i}(\boldsymbol{x})$: household consumption of agricultural products at location $\boldsymbol{x}$
$c_{3i}(\boldsymbol{x})$: household consumption of forestry products at location $\boldsymbol{x}$
$c_{4i}(\boldsymbol{x})$: household consumption of rural goods at location $\boldsymbol{x}$
$c_{Fi}(\boldsymbol{x})$: household future consumption at location $\boldsymbol{x}$
$m_{Hi}(\boldsymbol{x})$: household land input at location $\boldsymbol{x}$
n_{U0}: natural environmental level (e.g. air and water)
n_{R3}: forest volume in the rural area
$\beta_{k1}(k = 1, 2, 3, 4)$, β_{F1}, β_{m1}, β_{U01} and β_{R31}: elasticity parameters ($\beta_{11} + \beta_{21} + \beta_{31} + \beta_{41} + \beta_{F1} + \beta_{m1} + \beta_{U01} + \beta_{R31} = 1$)
c: expected damage rate on household asset ($c = 0$ in the normal land, $0 < c < 1$ in the flood-prone areas)
$Y_i(\boldsymbol{x})$: household income at location $\boldsymbol{x}$
$w_i(\boldsymbol{x})$: wage rate at location $\boldsymbol{x}$
$r_1(\boldsymbol{x})$: capital return rate at location $\boldsymbol{x}$
$ks_i(\boldsymbol{x})$: capital stock endowment of a household of type i at location $\boldsymbol{x}$
$\pi_{Hi}(\boldsymbol{x})$: redistributed income from the local government to the household at location $\boldsymbol{x}$

Each household endows available working time, l_{si}, and is assumed to perfectly, inelastically supply it to firms obtaining an income of $w_i(\boldsymbol{x})l_{si}$, plus redistributed income from the local government, $\pi_{Hi}(\boldsymbol{x})$. In household locational equilibrium, the utility levels of both types of households consider the same values u_i^*, irrespective of household locations. u_1^* is endogenously determined in the urban area, whereas u_2^* is endogenously determined in the rural area. Therefore, for current and future household consumption, the bid-max lot size and the bid-rent function in the two types of households are derived.

Moreover, future consumption becomes savings, which are then invested into the capital stock of each household. To explain this household behavior, first, derivation of the price of future goods is described. Future goods imply future consumption, derived from household savings. However, saving formulates capital investment. Therefore, capital goods can be regarded as savings goods. Investment is made by using only urban goods, q_1. Then, the price of investment goods is identified as $p_1(\boldsymbol{x})$. This can be regarded as the price of the savings goods, $p_s(\boldsymbol{x})$.

Because the capital returns from a unit of capital injection is equal to $r_1(\boldsymbol{x})$, the relationship of the expected return rate of the price of the savings good, $p_s(\boldsymbol{x})$, and the expected net return rate of household saving, $r_s(\boldsymbol{x})$, is written as $r_s(\boldsymbol{x}) = r_1(\boldsymbol{x})/p_s(\boldsymbol{x})$. It is assumed that the expected returns of savings will finance future consumption. Regarding the price of future goods as the price of the current consumption goods under the myopic expectation and denoting the household real saving by $s_i(\boldsymbol{x})$, the following equation holds.

$$p_{Gi}\,(\boldsymbol{x})\,c_{Fi}\,(\boldsymbol{x}) = r_1\,(\boldsymbol{x})\,s_i\,(\boldsymbol{x}) \tag{5.18}$$

$$p_{Gi}(\boldsymbol{x}) \equiv \prod_{k=1}^{4}\left[\frac{p_k(\boldsymbol{x})}{\beta_{k1}}\right]^{\frac{\beta_{k1}}{\beta_{T1}}}\left[\frac{p_F(\boldsymbol{x})}{\beta_{F1}}\right]^{\frac{\beta_{F1}}{\beta_{T1}}}\left[\frac{g(\boldsymbol{x})}{\beta_{m1}}\right]^{\frac{\beta_{m1}}{\beta_{T1}}}, \tag{5.19}$$

where $\beta_{T1} \equiv \sum_{k=1}^{4}\beta_{k1} + \beta_{F1} + \beta_{m1}$ and $g(\boldsymbol{x})$ is the market land rent at location $\boldsymbol{x}$.
This yields:

$$\left[\frac{p_s(\boldsymbol{x})\, p_{Gi}(\boldsymbol{x})}{r_1(\boldsymbol{x})}\right] c_{Fi}(\boldsymbol{x}) = p_s(\boldsymbol{x})\, s_i(\boldsymbol{x}),$$

and we set the price of future good $p_{Fi}(\boldsymbol{x})$, associated with the real saving $s_i(\boldsymbol{x})$, as:

$$p_{Fi}(\boldsymbol{x}) = \frac{p_s(\boldsymbol{x})\, p_{Gi}(\boldsymbol{x})}{r_1(x)}. \tag{5.20}$$

Then, $p_s(\boldsymbol{x})s_i(\boldsymbol{x}) = p_{Fi}(\boldsymbol{x})c_{Fi}(\boldsymbol{x})$ is realized.
Now the household bid-rent function for land is specified as follows.

$$g_{Hi}(\boldsymbol{x}) \equiv \max\left[\frac{1}{m_{Hi}(\boldsymbol{x})}\left\{w_i(\boldsymbol{x}) + r_1(\boldsymbol{x})\, k s_i(\boldsymbol{x}) + \pi_{Hi}(\boldsymbol{x}) - \sum_{k=1}^{4} p_k(\boldsymbol{x})\, c_{ki}(\boldsymbol{x}) - p_{Fi}(\boldsymbol{x})\, c_{Fi}(\boldsymbol{x})\right\}\right], \tag{5.21}$$

with respect to $c_{ki}(\boldsymbol{x})$, $c_{Fi}(\boldsymbol{x})$, and $m_{Hi}(\boldsymbol{x})$,
subject to:

$$u_i(\boldsymbol{x}) = u_i^*, \tag{5.22}$$

where $g_{Hi}(\boldsymbol{x})$ is household bid-rent function of type i at location $\boldsymbol{x}$ and $\pi_{Hi}(\boldsymbol{x})$ is defined as follows.

$$\pi_{Hi}(\boldsymbol{x}) \equiv \frac{\theta_i}{N}\iint_A \left[g_B(\boldsymbol{x})\, b(\boldsymbol{x})\, m_B(\boldsymbol{x}) + g_{H1}(\boldsymbol{x})\, h_1(\boldsymbol{x})\, m_{H1}(\boldsymbol{x}) + g_{H2}(\boldsymbol{x})\, h_2(\boldsymbol{x})\, m_{H2}(\boldsymbol{x})\right] dx_1 dx_2, \tag{5.23}$$

where $h_i(\boldsymbol{x})$ = density of households of type i at location $\boldsymbol{x}$ and $\theta_1 + \theta_2 = 1$. N is the number of households.

To solve the maximization problem Eqs. (5.21), (5.22), and (5.23), we consider the expenditure function.

$$E_i(\boldsymbol{x}) \equiv \min\left[\sum_{k=1}^{4} p_{ki}(\boldsymbol{x})\, c_{ki}(\boldsymbol{x}) + p_{Fi}(\boldsymbol{x})\, c_{Fi}(\boldsymbol{x}) + g_{Hi}(\boldsymbol{x})\, m_{Hi}(\boldsymbol{x})\right], \quad (i = 1, 2) \tag{5.24}$$

with respect to $c_{ki}(\boldsymbol{x})$, $c_{Fi}(\boldsymbol{x})$, and $m_{Hi}(\boldsymbol{x})$
subject to:

$$u_i^* = \frac{\prod_{k=1}^{4} c_{ki}(\boldsymbol{x})^{\beta_{k1}} c_{Fi}(\boldsymbol{x})^{\beta_{F1}} m_{Hi}(\boldsymbol{x})^{\beta_{m1}} (n_{U0})^{\beta_{U01}} (n_{R3})^{\beta_{R31}}}{1 + c(n_{R3})^{-\omega} Y_i(\boldsymbol{x})^{\varepsilon}} \quad (i = 1, 2). \tag{5.25}$$

The expenditure function is solved as follows:

$$E_i(\boldsymbol{x}) \equiv \beta_{T1} \left[\frac{\left\{1 + c(n_{R3})^{-\omega} Y_i(\boldsymbol{x})^{\varepsilon}\right\} u_i^*}{(n_{U0})^{\beta_{U01}} (n_{R3})^{\beta_{R31}}} \right] \times \prod_{k=1}^{4} \left[\frac{p_k(\boldsymbol{x})}{\beta_{k1}} \right]^{\frac{\beta_{k1}}{\beta_{T1}}} \left[\frac{p_{Fi}(\boldsymbol{x})}{\beta_{F1}} \right]^{\frac{\beta_{F1}}{\beta_{T1}}} \left[\frac{g_{Hi}(\boldsymbol{x})}{\beta_{m1}} \right]^{\frac{\beta_{m1}}{\beta_{T1}}}. \tag{5.26}$$

The expenditure function must be equal to the household income yielding the bid-rent function:

$$E_i(\boldsymbol{x}) \equiv \beta_{T1} \left[\frac{\left\{1 + c(n_{R3})^{-\omega} Y_i(\boldsymbol{x})^{\varepsilon}\right\} u_i^*}{(n_{U0})^{\beta_{U01}} (n_{R3})^{\beta_{R31}}} \right] \times \prod_{k=1}^{4} \left[\frac{p_k(\boldsymbol{x})}{\beta_{k1}} \right]^{\frac{\beta_{k1}}{\beta_{T1}}} \left[\frac{p_{Fi}(\boldsymbol{x})}{\beta_{F1}} \right]^{\frac{\beta_{F1}}{\beta_{T1}}} \left[\frac{g_{Hi}(\boldsymbol{x})}{\beta_{m1}} \right]^{\frac{\beta_{m1}}{\beta_{T1}}}, \tag{5.26}$$

$$g_{Hi}(\boldsymbol{x}) \equiv \frac{\beta_{m1}}{(\beta_{T1})^{\frac{\beta_{T1}}{\beta_{m1}}}} \left[\frac{(n_{U0})^{\beta_{U01}} (n_{R3})^{\beta_{R31}}}{\left\{1 + c(n_{R3})^{-\omega} Y_i(\boldsymbol{x})^{\varepsilon}\right\} u_i^*} \right] \times \prod_{k=1}^{4} \left[\frac{\beta_{k1}}{p_k(\boldsymbol{x})} \right]^{\frac{\beta_{k1}}{\beta_{m1}}} \left[\frac{\beta_{F1}}{p_{Fi}(\boldsymbol{x})} \right]^{\frac{\beta_{F1}}{\beta_{m1}}} Y_i(x)^{\frac{\beta_{T1}}{\beta_{m1}}}. \tag{5.27}$$

Then, demands for commodities, future goods, and the bid-max lot size are solved as follows:

$$c_{ki}(\boldsymbol{x}) = \frac{\beta_{k1}}{\beta_{T1} p_k(\boldsymbol{x})} Y_i(\boldsymbol{x}) \quad (k = 1, 2, 3, 4), \tag{5.28}$$

$$c_{Fi}(\boldsymbol{x}) = \frac{\beta_{F1}}{\beta_{T1} p_{Fi}(\boldsymbol{x})} Y_i(\boldsymbol{x}), \tag{5.29}$$

$$s_i(\boldsymbol{x}) = p_{Fi}(\boldsymbol{x})\, c_{Fi}(\boldsymbol{x}) / p_s(\boldsymbol{x}), \tag{5.30}$$

$$I_i(\boldsymbol{x}) = s_i(\boldsymbol{x}), \tag{5.31}$$

$$\begin{aligned} mH_i(\boldsymbol{x}) = & \left[\frac{\{1 + c(n_{R3})^{-\omega} Y_i(\boldsymbol{x})^{\varepsilon}\} u_i^*}{(n_{U0})^{\beta_{U01}} (n_{R3})^{\beta_{R31}}} \right]^{\beta_{T1}} \\ & \times \prod_{k=1}^{4} \left[\frac{\beta_{T1} p_k(\boldsymbol{x})}{\beta_{k1} Y_i(\boldsymbol{x})} \right]^{\frac{\beta_{k1}}{\beta_{m1}}} \left[\frac{\beta_{T1} p_{Fi}(\boldsymbol{x})}{\beta_{F1} Y_i(\boldsymbol{x})} \right]^{\frac{\beta_{F1}}{\beta_{m1}}}. \end{aligned} \tag{5.32}$$

Thus, the following dynamic equation holds.

$$\dot{ks}_i(\boldsymbol{x}) = I_i(\boldsymbol{x}) - \delta \cdot ks_i(\boldsymbol{x}) - \left(\frac{\dot{N}_i}{N_i} \right) ks_i(\boldsymbol{x}), \tag{5.33}$$

$$KS(t) \equiv \iint_A [ks_1(\boldsymbol{x}) h_1(\boldsymbol{x}) + ks_2(\boldsymbol{x}) h_2(\boldsymbol{x})] \, dx_1 dx_2, \tag{5.34}$$

where:

$ks_i(\boldsymbol{x})$: capital stock endowed by a household of type i at time t and location $\boldsymbol{x}$
$KS(t)$: aggregate capital stock at time t
$I_i(\boldsymbol{x})$: investment by a household of i at time t and location $\boldsymbol{x}$
δ: capital depreciation rate

Moreover, the natural environment in the city is specified as follows:

$$\begin{aligned} z_1 = & \iint_A \eta_1 q_1(\boldsymbol{x}) b(\boldsymbol{x}) \, dx_1 dx_2 \\ & + \iint_A \sum_{k=1}^{4} \sum_{i=1}^{2} \mu_{ki} c_{ki}(\boldsymbol{x}) h_i(\boldsymbol{x}) \, dx_1 dx_2, \end{aligned} \tag{5.35}$$

$$n_{U0} = \overline{n}_{U0} - \frac{z_1}{q_{A5} (m_{A3})^{\gamma_{B5}} (n_{R3})^{\gamma_{R5}}}, \tag{5.36}$$

where:

z_1: pollution
η_1, μ_{ki}: emission factors
$\overline{n}_{U0}$: natural environment before pollution
m_{A3}: agricultural land size in the rural area
n_{R3}: forest volume in the rural area

5.5 Firms' Behavior in Rural Areas

We consider an aggregate production function in respective sectors in the rural area. The production functions are specified as follows:

$$q_2 = q_{A2}\, q_{12}^{\alpha_{12}} q_{22}^{\alpha_{22}} q_{32}^{\alpha_{32}} q_{42}^{\alpha_{42}} ld_2^{\alpha_{l2}} kd_2^{\alpha_{k2}} m_{B2}^{\alpha_{m2}} n_{R0}^{\alpha_{R02}} n_{R3}^{\alpha_{R32}}, \tag{5.37}$$

$$q_3 = \min\left[q_{A3} q_{13}^{\alpha_{13}} q_{23}^{\alpha_{23}} q_{33}^{\alpha_{33}} q_{43}^{\alpha_{43}} ld_3^{\alpha_{l3}} kd_3^{\alpha_{k3}}, \frac{y_3}{\alpha_{y3}} \right], \tag{5.38}$$

$$q_4 = q_{A4}\, q_{14}^{\alpha_{14}} q_{24}^{\alpha_{24}} q_{34}^{\alpha_{34}} q_{44}^{\alpha_{44}} ld_4^{\alpha_{l4}} kd_4^{\alpha_{k4}} m_{B4}^{\alpha_{m4}} n_{R0}^{\alpha_{R04}} n_{R3}^{\alpha_{R34}}, \tag{5.39}$$

where:

$$\alpha_{12} + \alpha_{22} + \alpha_{32} + \alpha_{42} + \alpha_{l2} + \alpha_{k2} + \alpha_{m2} = 1$$

$$\alpha_{13} + \alpha_{23} + \alpha_{33} + \alpha_{43} + \alpha_{l3} + \alpha_{k3} = 1$$

$$\alpha_{14} + \alpha_{24} + \alpha_{34} + \alpha_{44} + \alpha_{l4} + \alpha_{k4} + \alpha_{m4} = 1$$

q_2: agricultural output
q_3: forestry output
q_4: output of rural general goods
q_{ki}: intermediate input
ld_i: labor input
kd_i: capital input
m_{B2}: agricultural land
m_{B4}: land input in rural general firm
y: cut volume of forest
n_{R0}: natural environmental level in the rural area
n_{R3}: forest volume in the rural area
q_{Ai}: efficient parameter
α_{ji}, α_{li}, α_{ki}, α_{mi}, α_{R0i}, and α_{R3i}: elasticity parameters
α_{y3}: Leontief parameter

The natural environment and the forest volume in the production functions are treated as externalities. Thus, the production functions are homogeneous of degree unity with respect to factor inputs. Then, we consider cost minimization in firm's behavior caused by linear homogeneity of degree, leading to the conditional demands for intermediate goods, labor, capital, land, and volume of forest cut.

$$q_{ij} = \left[\frac{q_j\alpha_{ij}}{q_{Aj}(n_{Ro})^{\alpha_{Roj}}(n_{R3})^{\alpha_{R3j}}p_i}\right] \times \prod_{i=1}^{4}\left[\frac{p_i}{\alpha_{ij}}\right]^{\alpha_{ij}}\left[\frac{w_3}{\alpha_{lj}}\right]^{\alpha_{lj}}\left[\frac{r_3}{\alpha_{kj}}\right]^{\alpha_{kj}}\left[\frac{g_{Bj}}{\alpha_{mj}}\right]^{\alpha_{mj}} \quad (j = 2, 4), \tag{5.40}$$

$$q_{i3} = \left[\frac{q_3\alpha_{i3}}{q_{A3}p_i}\right]\prod_{i=1}^{4}\left[\frac{p_i}{\alpha_{i3}}\right]^{\alpha_{ij}}\left[\frac{w_3}{\alpha_{l3}}\right]^{\alpha_{ij}}\left[\frac{r_3}{\alpha_{k3}}\right]^{\alpha_{kj}}\left[\frac{g_{B3}}{\alpha_{m3}}\right]^{\alpha_{mj}}, \tag{5.41}$$

$$ld_j = \left[\frac{q_j\alpha_{lj}}{q_{Aj}(n_{Ro})^{\alpha_{Roj}}(n_{R3})^{\alpha_{R3j}}w_3}\right] \times \prod_{i=1}^{4}\left[\frac{p_i}{\alpha_{ij}}\right]^{\alpha_{ij}}\left[\frac{w_3}{\alpha_{lj}}\right]^{\alpha_{lj}}\left[\frac{r_3}{\alpha_{kj}}\right]^{\alpha_{kj}}\left[\frac{g_{Bj}}{\alpha_{mj}}\right]^{\alpha_{mj}} \quad (j = 2, 4), \tag{5.42}$$

$$ld_3 = \left[\frac{q_3\alpha_{l3}}{q_{A3}w_3}\right]\prod_{i=1}^{4}\left[\frac{p_i}{\alpha_{i3}}\right]^{\alpha_{i3}}\left[\frac{w_3}{\alpha_{l3}}\right]^{\alpha_{l3}}\left[\frac{r_3}{\alpha_{k3}}\right]^{\alpha_{k3}}\left[\frac{g_{B3}}{\alpha_{m3}}\right]^{\alpha_{m3}}, \tag{5.43}$$

$$kd_j = \left[\frac{q_j\alpha_{kj}}{q_{Aj}(n_{Ro})^{\alpha_{Roj}}(n_{R3})^{\alpha_{R3j}}r_3}\right] \times \prod_{i=1}^{4}\left[\frac{p_i}{\alpha_{ij}}\right]^{\alpha_{ij}}\left[\frac{w_3}{\alpha_{lj}}\right]^{\alpha_{lj}}\left[\frac{r_3}{\alpha_{kj}}\right]^{\alpha_{kj}}\left[\frac{g_{Bj}}{\alpha_{mj}}\right]^{\alpha_{mj}} \quad (j = 2, 4), \tag{5.44}$$

$$kd_3 = \left[\frac{q_3\alpha_{k3}}{q_{A3}r_3}\right]\prod_{i=1}^{4}\left[\frac{p_i}{\alpha_{i3}}\right]^{\alpha_{i3}}\left[\frac{w_3}{\alpha_{l3}}\right]^{\alpha_{l3}}\left[\frac{r_3}{\alpha_{k3}}\right]^{\alpha_{k3}}\left[\frac{g_{B3}}{\alpha_{m3}}\right]^{\alpha_{m3}}, \tag{5.45}$$

$$m_{Bj} = \left[\frac{q_j\alpha_{mj}}{q_{Aj}(n_{Ro})^{\alpha_{Roj}}(n_{R3})^{\alpha_{R3j}}g_{Bj}}\right] \times \prod_{i=1}^{4}\left[\frac{p_i}{\alpha_{ij}}\right]^{\alpha_{ij}}\left[\frac{w_3}{\alpha_{lJ}}\right]^{\alpha_{lj}}\left[\frac{r_3}{\alpha_{kJ}}\right]^{\alpha_{kj}}\left[\frac{g_{Bj}}{\alpha_{mj}}\right]^{\alpha_{mj}} \quad (j = 2, 4), \tag{5.46}$$

$$y_3 = \alpha_{y3}q_3. \tag{5.47}$$

5.6 Households' Behavior in Rural Areas

In the rural area, we do not consider the size and location of the area. Thus, we specify household behavior as intertemporal utility maximization. The representative household behavior may be illustrated as follows:

$$\max\int_0^\infty \frac{\sigma_3}{\sigma_3 - 1}\left[\prod_{k=1}^{4} c_{k3}^{\beta_{k3}} m_{H3}^{\beta_{m3}} n_{R0}^{\beta_{R03}} n_{R3}^{\beta_{R33}}\right]^{\frac{\sigma_3 - 1}{\sigma_3}} e^{-\xi t}dt. \tag{5.48}$$

subject to:

$$\dot{ks}_3 = \left[w_3 + (r_3 - p_1(\boldsymbol{x}_B)\,\delta)\,ks_3 + g_{B2}ms_{B2} + g_{B4}ms_{B4} + g_{H3}ms_{H3} - \sum_{k=1}^{4} p_k c_{k3} - g_{H3}m_{H3} \right] \frac{1}{p_1(\boldsymbol{x}_B)} - \left(\frac{\dot{N}_3}{N_3}\right) ks_3, \tag{5.49}$$

$$I_3 \equiv \left[w_3 + r_3 ks_3 + g_{B2}ms_{B2} + g_{B4}ms_{B4} + g_{H3}ms_{H3} - \sum_{k=1}^{4} p_k c_{k3} - g_{H3}m_{H3} \right] \frac{1}{p_1(\boldsymbol{x}_B)}, \tag{5.50}$$

where:

σ_3: intertemporal elasticity of substitution
c_{13}: household consumption of urban goods
c_{23}: household consumption of agricultural products
c_{33}: household consumption of forestry products
c_{43}: household consumption of rural goods
m_{H3}: residential land
ξ: subjective discount rate
ks_3: capital stock endowed by a household
w_3: wage rate prevailing in the rural area
r_3: capital return rate prevailing in the rural area
δ: capital depreciation rate
p_i: price of good i
$g_{H3}3$: residential land rent
$\dot{N}_3/N_3$: population growth rate in the rural area

Moreover, we should consider the natural environment and forests. Those are specified as follows:

$$z_3 = \sum_{k=2}^{4} \eta_k q_k + \sum_{k=2}^{4} \mu_k N_3 c_{k3}, \tag{5.51}$$

$$n_{R0} = \overline{n}_{R0} - \frac{z_3}{q_{B6}(m_{B2})^{\gamma_{B6}}(n_{R3})^{\gamma_{R6}}}, \tag{5.52}$$

$$\dot{n}_{R3} = (\varepsilon_3 - \theta_3 n_{\mathrm{R3}})\, n_{R3} - y_3, \tag{5.53}$$

where:

z_3: pollution in the rural area
η_k and μ_k: emission factors

$\overline{n}_{R0}$: natural environmental level before pollution
ε_3: carrying capacity
θ_3: ecological parameter

To solve the rural household optimization behavior, we introduce the current value Hamiltonian, which is:

$$H_3 \equiv \int_0^{\infty} \frac{\sigma_3}{\sigma_3 - 1} \left[\prod_{k=1}^{4} c_{k3}^{\beta_{k3}} m_{H3}^{\beta_{m3}} n_{R0}^{\beta_{R03}} n_{R3}^{\beta_{R33}} \right]^{\frac{\sigma_3 - 1}{\sigma_3}}$$
$$+\lambda_3 \left\{ \left[w_3 + (r_3 - p_1(\boldsymbol{x}_B)\delta) ks_3 + g_{B2} ms_{B2} + g_{B4} ms_{B4} + g_{H3} ms_{H3} \right.\right. \tag{5.54}$$
$$\left.\left. - \sum_{k=1}^{4} p_k c_{k3} - g_{H3} m_{H3} \right] \frac{1}{p_1(\boldsymbol{x}_B)} - \left(\frac{\dot{N}_3}{N_3} \right) ks_3 \right\}.$$

Applying the optimal control theory, we obtain the current consumption of goods, residential land, and the dynamic equation for the costate variable. The necessary and sufficient conditions for optimality are as follows:

$$\dot{\lambda}_3 = -\frac{\partial H_3}{\partial ks_3} + \xi \lambda_3. \tag{5.55}$$

c_{12}, c_{22}, c_{32}, c_{42}, and m_{H2} maximize the Hamiltonian at each time. The transversality condition is given as:

$$\lim_{t \to \infty} \lambda_3 \cdot \dot{ks_3} \cdot e^{-\xi t} = 0. \tag{5.56}$$

Those conditions are expressed as follows:

$$\dot{ks_3} = \left[w_3 + (r_3 - p_1(\boldsymbol{x}_B)\delta) ks_3 + g_{B2} ms_{B2} + g_{B4} ms_{B4} \right.$$
$$+ g_{H3} ms_{H3} - p_1(\boldsymbol{x}_B) c_{13} \tag{5.57}$$
$$\left. - \sum_{k=2}^{4} p_k c_{k3} - g_{H3} m_{H3} \right] \frac{1}{p_1(\boldsymbol{x}_B)} - \left(\frac{\dot{N}_3}{N_3} \right) ks_3,$$

$$\dot{\lambda}_3 = \lambda_3 \left[\xi - \frac{r_3}{p_1(\boldsymbol{x}_B)} + \delta + \frac{\dot{N}_3}{N_3} \right], \tag{5.58}$$

$$c_{i3} = \left[\frac{p_1(\boldsymbol{x}_B)}{\lambda_3} \right]^{\Theta_1} \left[\frac{\beta_{i3}}{p_i(\boldsymbol{x}_B)} \right] \prod_{k=1}^{4} \left[\frac{\beta_{k3}}{p_k} \right]^{\Theta_2} \left[\frac{\beta_{m3}}{g_{H3}} \right]^{\Theta_3} [n_{R0}]^{\Theta_4} [n_{R3}]^{\Theta_5}, \tag{5.59}$$

$$m_{H3} = \left[\frac{p_1(\boldsymbol{x}_B)}{\lambda_3} \right]^{\Theta_1} \left[\frac{\beta_{m3}}{g_{H3}} \right] \prod_{k=1}^{4} \left[\frac{\beta_{k3}}{p_k} \right]^{\Theta_2} \left[\frac{\beta_{m3}}{g_{H3}} \right]^{\Theta_3} [n_{R0}]^{\Theta_4} [n_{R3}]^{\Theta_5}, \tag{5.60}$$

where:

$\Theta_1 \equiv \frac{\sigma_3}{\sigma_3+(1-\sigma_3)\beta_{T3}}$
$\Theta_2 \equiv \frac{(\sigma_3-1)\beta_{k3}}{\sigma_3+(1-\sigma_3)\beta_{T3}}$
$\Theta_3 \equiv \frac{(\sigma_3-1)\beta_{m3}}{\sigma_3+(1-\sigma_3)\beta_{T3}}$
$\Theta_4 \equiv \frac{\sigma_3\beta_{R03}}{\sigma_3+(1-\sigma_3)\beta_{T3}}$
$\Theta_5 \equiv \frac{\sigma_3\beta_{R33}}{\sigma_3+(1-\sigma_3)\beta_{T3}}$
$\beta_{T3} \equiv \beta_{13} + \beta_{23} + \beta_{33} + \beta_{43} + \beta_{m3}.$

5.7 Property of the Hamiltonian

Let us further examine the implication of the Hamiltonian. Along the optimal trajectory, all variables in the model can be represented by functions of stock variables and their associated costate variables. Thus, the time derivative in the Hamiltonian on the optimal path, H^*, can be calculated by applying the canonical system of equations of the stock variables:

$$\begin{aligned} \frac{dH_3^*}{dt} &= \frac{\partial H_3^*}{\partial k_3}\frac{dk_3}{dt} + \frac{\partial H_3^*}{\partial \lambda_3}\frac{d\lambda_3}{dt} \\ &= \left(\xi\lambda_3 - \frac{d\lambda_3}{dt}\right)\frac{dk_3}{dt} + \frac{dk_3}{dt}\frac{d\lambda_3}{dt} \\ &= \xi\lambda_3\frac{dk_3}{dt} = \xi\left(H_3^* - u_3^*\right), \end{aligned} \tag{5.61}$$

where u_3^* is the value of the utility function on the optimal trajectory. Solving this differential equation, one can obtain another expression of the Hamiltonian:

$$H_3^*(t) = \xi \int_t^\infty u_3^*(\tau)\, e^{-\xi(\tau-t)} d\tau. \tag{5.62}$$

That is, the value of the Hamiltonian at time t, $H_3^*(t)$), is calculated as the integration of the present value of the maximized utility multiplied by the subjective discount rate, ξ. Furthermore, the calculation on $H_3^*(t)$ yields:

$$\int_t^\infty H_3^*(t)e^{-\xi(\tau-t)} d\tau = \int_t^\infty u_3^*(\tau)\, e^{-\xi(\tau-t)} d\tau. \tag{5.63}$$

Equation (5.63), in turn, indicates that the integration of discounted constant income stream, $H_3^*(t)$, equals the integration of the present value of the household utility. Therefore, $H_3^*(t)$ can be regarded as a social welfare index, giving the stationary equivalent to the future utility.

5.8 Land Market Equilibrium Conditions

Denoting the agricultural land rent as g_{B2}, determined in the rural area, the market rent function, $g(\boldsymbol{x})$, over the city in equilibrium, is described as follows:

On the normal land:

$$g(\boldsymbol{x}) \equiv \max\left\{g_B(\boldsymbol{x}), g_{H1}(\boldsymbol{x}), g_{H2}(\boldsymbol{x}), g_{B2}\right\}. \tag{5.64}$$

In the flood-prone area:

$$g(\boldsymbol{x}) \equiv \max\left\{g_B(\boldsymbol{x}), g_{H1}(\boldsymbol{x}), g_{H2}(\boldsymbol{x})\right\}. \tag{5.65}$$

The reason for Eq. (5.65) is that the flood-prone area cannot be used for agriculture because of its periodic inundation. The flood-prone area is assumed to be located within the residential area. When we assume that the business area is located around the city center, and the residential area is located outside the business area, the land equilibrium conditions are expressed as follows (Miyata 2011):

$$g(\boldsymbol{x}) = g_B(\boldsymbol{x}) \geq g_{H1}(\boldsymbol{x}) \text{ and } g_{H2}(\boldsymbol{x}) \text{ for } \boldsymbol{x} \in \text{ business area,} \tag{5.66}$$

$$g(\boldsymbol{x}) = g_{B1}(\boldsymbol{x}) = g_{H1}(\boldsymbol{x}) \text{ for } \boldsymbol{x} \in \text{ the boundary between the business and the residential areas,} \tag{5.67}$$

$$g(\boldsymbol{x}) = g_{H1}(\boldsymbol{x}) \geq g_{B1}(\boldsymbol{x}) \text{ and } g_{H2}(\boldsymbol{x}) \text{ for } \boldsymbol{x} \in \text{ residential area on normal land,} \tag{5.68}$$

$$g(\boldsymbol{x}) = g_{H2}(\boldsymbol{x}) \geq g_{B1}(\boldsymbol{x}) \text{ and } g_{H1}(\boldsymbol{x}) \text{ for } \boldsymbol{x} \in \text{ flood} - \text{prone area,} \tag{5.69}$$

$$g(\boldsymbol{x}) = g_{H1}(\boldsymbol{x}) = g_{B2} \text{ for } \boldsymbol{x} \in \text{ city boundary,} \tag{5.70}$$

$$N_3 ms_{B2} = m_{B2} \text{ agricultural land equilibrium condition in the rural area,} \tag{5.71}$$

$$N_3 ms_{B4} = m_{B4} \text{ firms land equilibrium condition in the rural area, and} \tag{5.72}$$

$$N_3 ms_{H3} = N_3 m_{H3} \text{ residential land equilibrium condition in the rural area.} \tag{5.73}$$

5.9 Local Balance Equations for Commodity and Labor

Let us assume the transport technology for commodities is a von Thünen technology with a cost ratio, a_i. That is, the transport cost is incurred in the transported commodities, expressed as $a_i q_i$ for carrying q_i units of commodities for the unit distance. Let $\varphi_i(\boldsymbol{x})$ be a 2D vector of commodities of type i (e.g., urban, agricultural, forestry and rural) transported to location $\boldsymbol{x}$ in a unit time and of a unit area size. Then, the following local balance equations for commodities hold (Beckmann 1952; Beckmann and Puu 1985; and Puu 2003):

$$\begin{aligned} \operatorname{div} \varphi_1(\boldsymbol{x}) = q_1(\boldsymbol{x})\,\mathrm{b}(\boldsymbol{x}) - I_1(\boldsymbol{x})\,b(\boldsymbol{x}) - c_{11}(\boldsymbol{x})\,h_1(\boldsymbol{x}) - c_{12}(\boldsymbol{x})\,h_2(\boldsymbol{x}) \\ - a_1 \left\|\varphi_1(\boldsymbol{x})\right\|, \end{aligned} \tag{5.74}$$

$$\begin{aligned} \operatorname{div} \varphi_2(\boldsymbol{x}) = -b(\boldsymbol{x})\,q_{21}(\boldsymbol{x}) - c_{21}(\boldsymbol{x})\,h_1(\boldsymbol{x}) - c_{22}(\boldsymbol{x})\,h_2(\boldsymbol{x}) \\ - a_2 \left\|\varphi_2(\boldsymbol{x})\right\|, \end{aligned} \tag{5.75}$$

$$\begin{aligned} \operatorname{div} \varphi_3(\boldsymbol{x}) = -b(\boldsymbol{x})\,q_{31}(\boldsymbol{x}) - c_{31}(\boldsymbol{x})\,h_1(\boldsymbol{x}) - c_{32}(\boldsymbol{x})\,h_2(\boldsymbol{x}) \\ - a_3 \left\|\varphi_3(\boldsymbol{x})\right\|, \end{aligned} \tag{5.76}$$

$$\begin{aligned} \operatorname{div} \varphi_4(\boldsymbol{x}) = -b(\boldsymbol{x})\,q_{41}(\boldsymbol{x}) - c_{41}(\boldsymbol{x})\,h_1(\boldsymbol{x}) - c_{42}(\boldsymbol{x})\,h_2(\boldsymbol{x}) \\ - a_4 \left\|\varphi_4(\boldsymbol{x})\right\|, \end{aligned} \tag{5.77}$$

where:

$b(\boldsymbol{x})$: density of firms at $\boldsymbol{x}$ in the city area
$h_1(\boldsymbol{x})$: density of H.I.H.s at $\boldsymbol{x}$ in the city
$h_2(\boldsymbol{x})$: density of L.I.H.s at $\boldsymbol{x}$ in the city
a_i: von Thünen coefficient
$\|\cdot\|$: norm of a 2D vector

Similarly, let b_i denote the transport cost of transporting unit labor of H.I.H.s or L.I.H.s at a unit distance. This cost is also incurred as labor. Let $\psi_i(\boldsymbol{x})$ express a 2D vector of labor in unit time and in unit area size, transported to location $\boldsymbol{x}$. Then, the local balance equation for labor held at location $\boldsymbol{x}$ is expressed as follows (Beckmann and Puu 1985):

$$\operatorname{div} \psi_i(\boldsymbol{x}) = ls_i h_i(\boldsymbol{x}) - ld_i(\boldsymbol{x})\,b(\boldsymbol{x}) - b_i \left\|\psi_i(\boldsymbol{x})\right\| \; (i = 1, 2). \tag{5.78}$$

5.10 The Local Government and Absentee Landowners

The city area is assumed to be occupied by absentee landowners, who reside outside of the city. The local government rents all land in the city from the absentee landowners. It rents the land to firms and households in the city at market rate rents. The revenue of the local government from the land rent is redistributed to households. The redistributed household income, $\pi_{Hi}(\boldsymbol{x})$, is indicated in Eq. (5.23). $\theta_1(\boldsymbol{x})$ and $\theta_2(\boldsymbol{x})$ in Eq. (5.23) are policy parameters that aim to reduce illegal settlements in the flood-prone area.

5.11 Global Equilibrium Condition in Commodity and Labor Markets

Integrating the commodity local balance Eq. (5.74), and applying the Gauss divergence theorem, one can obtain the global equilibrium condition on commodities as presented in Eq. (5.79):

$$\begin{aligned}\iint_A \operatorname{div} \varphi_1(\boldsymbol{x})\, dx_1 dx_2 &= \iint_A [q_1(\boldsymbol{x})\, b(\boldsymbol{x}) - I_1(\boldsymbol{x})\, b_1(\boldsymbol{x}) - c_{11}(\boldsymbol{x})\, h_1(\boldsymbol{x}) \\ &\quad - c_{12}(\boldsymbol{x})\, h_2(\boldsymbol{x}) - a_1 \| \varphi_1(\boldsymbol{x}) \|]\, dx_1 dx_2 \\ &= \int_{\partial A} q_{n1}(\boldsymbol{x}(s))\, ds = q_{12} + q_{13} + q_{14} + c_{13} + I_3. \end{aligned} \tag{5.79}$$

Next, integrating the labor local balance Eq. (5.78) with the Gauss divergence theorem, one can obtain the global equilibrium equation for labor. As a result of the face that we assume that there is no in- and out-migration at the city boundary at equilibrium, the line integral of labor migration along the city boundary also becomes zero.

$$\begin{aligned}\iint_A \operatorname{div} \psi_i(\boldsymbol{x})\, dx_1 dx_2 &= \iint_A [ls_i h_i(\boldsymbol{x}) - ld_i(\boldsymbol{x})\, b_1(\boldsymbol{x}) - b_i \|\psi_i(\boldsymbol{x})\|]\, dx_1 dx_2 \\ &= \int_{\partial A} q_{n1}(\boldsymbol{x}(s))\, ds = 0 \ (i = 1, 2), \end{aligned} \tag{5.80}$$

5.12 Commodity Transport Agent (C.T.A.) and Labor Transport Agents (L.T.A1. and L.T.A2.)

The transportation of commodities is assumed to be performed by the commodity transport agent (C.T.A.) (Beckmann 1952; Beckmann and Puu 1985; and Puu 2003). The C.T.A. buys $p_1(\boldsymbol{x})q_1(\boldsymbol{x})b_1(\boldsymbol{x})$ of commodities at point $\boldsymbol{x}$ and sells

$p_1(\boldsymbol{x})c_1(\boldsymbol{x})h_1(\boldsymbol{x}) + p_1(\boldsymbol{x})c_2(\boldsymbol{x})h_2(\boldsymbol{x})$ of commodities to households. Thus, the profit of C.T.A. at $\boldsymbol{x}$ is expressed as follows:

$$\begin{aligned} \pi_{Tq1}(\boldsymbol{x}) &= p_1(\boldsymbol{x})\, c_{11}(\boldsymbol{x})\, h_1(\boldsymbol{x}) + p_1(\boldsymbol{x})\, c_{21}(\boldsymbol{x})\, h_2(\boldsymbol{x}) \\ &\quad + p_1(\boldsymbol{x})\, I_1(\boldsymbol{x})\, b(\boldsymbol{x}) - p_1(\boldsymbol{x})\, q_1(\boldsymbol{x})\, b(\boldsymbol{x}) \\ &= -p_1(\boldsymbol{x})\, \mathrm{div}\, \varphi_1(\boldsymbol{x}) - p_1(\boldsymbol{x})\, a_1 \|\varphi_1(\boldsymbol{x})\|, \end{aligned} \tag{5.81}$$

$$\begin{aligned} \pi_{Tq1}(\boldsymbol{x}_B) &= p_1(\boldsymbol{x}_B)\, c_{11}(\boldsymbol{x}_B)\, h_1(\boldsymbol{x}_B) + p_1(\boldsymbol{x}_B)\, c_{21}(\boldsymbol{x}_B)\, h_2(\boldsymbol{x}_B) \\ &\quad + p_1(\boldsymbol{x}_B)\, I_1(\boldsymbol{x}_B)\, b(\boldsymbol{x}_B) + p_1(\boldsymbol{x}_B)\, q_{12}(\boldsymbol{x}_B) + p_1(\boldsymbol{x}_B)\, q_{13}(\boldsymbol{x}_B) \\ &\quad + p_1(\boldsymbol{x}_B)\, q_{14}(\boldsymbol{x}_B) + p_1(\boldsymbol{x}_B)\, c_{13}(\boldsymbol{x}_B) + p_1(\boldsymbol{x}_B)\, I_3(\boldsymbol{x}_B) \\ &\quad - p_1(\boldsymbol{x}_B)\, q_1(\boldsymbol{x}_B)\, b(\boldsymbol{x}_B) = -p_1(\boldsymbol{x}_B)\, \mathrm{div}\, \varphi_1(\boldsymbol{x}_B) - p_1(\boldsymbol{x}_B)\, a_1 \|\varphi_1(\boldsymbol{x}_B)\|. \end{aligned} \tag{5.82}$$

The C.T.A. aims to find the optimal route that maximizes profit earned over the entire city area. The profit of the C.T.A. over the entire city area is written as follows:

$$\begin{aligned} \iint_A \pi_{Tq1}(\boldsymbol{x})\, dx_1 dx_2 &= \iint_A [p_1(\boldsymbol{x})\, c_{11}(\boldsymbol{x})\, h_1(\boldsymbol{x}) + p_1(\boldsymbol{x})\, c_{21}(\boldsymbol{x})\, h_2(\boldsymbol{x}) \\ &\quad + p_1(\boldsymbol{x})\, I_1(\boldsymbol{x})\, b(\boldsymbol{x}) - p_1(\boldsymbol{x})\, q_1(\boldsymbol{x})\, b(\boldsymbol{x})]\, dx_1 dx_2 \\ &\quad + \int_{\partial A} [p_1(\boldsymbol{x}_B(s))\, q_{12}(\boldsymbol{x}_B(s)) + p_1(\boldsymbol{x}_B(s))\, q_{13}(\boldsymbol{x}_B(s)) \\ &\quad + p_1(\boldsymbol{x}_B(s))\, q_{14}(\boldsymbol{x}_B(s)) + p_1(\boldsymbol{x}_B(s))\, c_{13}(\boldsymbol{x}_B(s)) \\ &\quad + p_1(\boldsymbol{x}_B(s))\, I_3(\boldsymbol{x}_B(s)) - p_1(\boldsymbol{x}_B(s))\, q_1(\boldsymbol{x}_B(s))\, b(\boldsymbol{x}_B(s))]\, ds \\ &= -\iint_A [p_1(\boldsymbol{x})\, \mathrm{div}\, \varphi_1(\boldsymbol{x}) + p_1(\boldsymbol{x})\, a_1 \|\varphi_1(\boldsymbol{x})\|]\, dx_1 dx_2 \\ &\quad + \int_{\partial A} [p_1(\boldsymbol{x}_B(s))\, q_{12}(\boldsymbol{x}_B(s)) + p_1(\boldsymbol{x}_B(s))\, q_{13}(\boldsymbol{x}_B(s)) \\ &\quad + p_1(\boldsymbol{x}_B(s))\, q_{14}(\boldsymbol{x}_B(s)) + p_1(\boldsymbol{x}_B(s))\, c_{13}(\boldsymbol{x}_B(s)) \\ &\quad + p_1(\boldsymbol{x}_B(s))\, I_3(\boldsymbol{x}_B(s)) - p_1(\boldsymbol{x}_B(s))\, q_1(\boldsymbol{x}_B(s))\, b(\boldsymbol{x}_B(s))]\, ds. \end{aligned} \tag{5.83}$$

The necessary and sufficient condition for the profit maximization in the C.T.A. is derived from the calculus variation (Gelfand and Fomin 1963). The Euler–Lagrange equation for the calculus variation is as follows:

$$\frac{d}{dx_1}\frac{\partial \pi_{Tq1}}{\partial\left(\frac{\partial \varphi_{1j}}{\partial x_1}\right)} + \frac{d}{dx_2}\frac{\partial \pi_{Tq1}}{\partial\left(\frac{\partial \varphi_{1j}}{\partial x_2}\right)} - \frac{\partial \pi_{Tq1}}{\partial \varphi_{1j}} = 0, \tag{5.84}$$

where $\boldsymbol{x} = (x_1, x_2)$ and $\varphi_1(\boldsymbol{x}) = (\varphi_{11}(x_1, x_2), \varphi_{12}(x_1, x_2))$.

Transforming the profit in the C.T.A. in the $x_1 - x_2$ coordinate, we have

$$\begin{aligned} \pi_{Tq1}(\boldsymbol{x}) &= -p_1(\boldsymbol{x})\, \mathrm{div}\, \varphi_1(\boldsymbol{x}) - p_1(\boldsymbol{x})\, a_1 \left|\left|\varphi_1(\boldsymbol{x})\right|\right| \\ &= -p_1(x_1, x_2)\left\{\frac{\partial \varphi_{11}}{\partial x_1} + \frac{\partial \varphi_{12}}{\partial x_2} + a_1\left(\varphi_{11}^2 + \varphi_{12}^2\right)^{\frac{1}{2}}\right\}. \end{aligned} \tag{5.85}$$

Therefore, the Euler–Lagrange equation is concretely expressed as follows:

$$\frac{\partial \pi_{Tq1}}{\partial \left(\frac{\partial \varphi_{1j}}{\partial x_i}\right)} = -p_1(\boldsymbol{x}), \tag{5.86}$$

$$\frac{d}{dx_i}\frac{\partial \pi_{Tq1}}{\partial \left(\frac{\partial \varphi_{1j}}{\partial x_i}\right)} = -\frac{\partial p_1(\boldsymbol{x})}{\partial x_i}, \tag{5.87}$$

$$\frac{\partial \pi_{Tq1}}{\partial \varphi_{1i}} = -p_1(\boldsymbol{x})\, a_1 \frac{\varphi_{1i}(\boldsymbol{x})}{||\varphi_1(\boldsymbol{x})||}, \tag{5.88}$$

$$\therefore p_1(\boldsymbol{x})\, a_1 \frac{\varphi_1(\boldsymbol{x})}{\|\varphi_1(\boldsymbol{x})\|} = \text{grad } p_1(\boldsymbol{x}). \tag{5.89}$$

In Eq. (5.89), $\varphi_1(\boldsymbol{x})/\|\varphi_1(\boldsymbol{x})\|$ is the direction along which commodities are transported, and that direction coincides the gradient of commodity price. This condition is optimal for commodity transport. Thus, the direction along which commodities are carried is the direction where commodity price becomes highest.

$p_1(\boldsymbol{x})a_1\varphi_1(\boldsymbol{x})/\|\varphi_1(\boldsymbol{x})\|$ depicts the cost of carrying a unit commodity in a unit distance. Let us calculate the transport cost of carrying a unit commodity from point $\boldsymbol{x}_A$ to $\boldsymbol{x}_B$ on the optimal route. We denote the route by:

$$D(s) = (x_1(s), x_2(s))$$

$$(0 \le s \le 1, \boldsymbol{x}_A = (x_1(0), x_2(0)), \boldsymbol{x}_B = (x_1(1), x_2(1))).$$

Thus, the transport cost is expressed as follows:

$$\begin{aligned} \int_0^1 p_1(\boldsymbol{x}(s))\, a_1 \tfrac{\varphi_1(\boldsymbol{x})}{\|\varphi_1(\boldsymbol{x})\|} \tfrac{d\boldsymbol{x}(s)}{ds} ds &= \int_0^1 \text{grad } p_1(\boldsymbol{x}(s)) \tfrac{d\boldsymbol{x}(s)}{ds} ds \\ &= p_1(\boldsymbol{x}_B) - p_1(\boldsymbol{x}_A). \end{aligned} \tag{5.90}$$

This equation implies that the transport cost of transporting the unit commodity on the optimal route is the difference between commodity prices at different points. This also asserts that the transport cost is incurred by commodity price. Let us calculate the profit the C.T.A. earned in the entire city area. Multiplying both sides of Eq. (5.89) by $\varphi_1(\boldsymbol{x})$ as a scalar product, we get:

$$p_1(\boldsymbol{x})\, a_1\, ||\varphi_1(\boldsymbol{x})|| = \varphi_1(\boldsymbol{x}) \text{ grad } p_1(\boldsymbol{x}), \tag{5.91}$$

$$
\begin{aligned}
\therefore \iint_A \pi_{Tq1}(\boldsymbol{x})\,dx_1dx_2 = &-\iint_A [p_1(\boldsymbol{x})\,\mathrm{div}\,\varphi_1(\boldsymbol{x}) + p_1(\boldsymbol{x})\,a_1\,\|\varphi_1(\boldsymbol{x})\|]\,dx_1dx_2 \\
&+ \int_{\partial A} [p_1(\boldsymbol{x}_B(s))\,q_{12}(\boldsymbol{x}_B(s)) + p_1(\boldsymbol{x}_B(s))\,q_{13}(\boldsymbol{x}_B(s)) \\
&+ p_1(\boldsymbol{x}_B(s))\,q_{14}(\boldsymbol{x}_B(s)) + p_1(\boldsymbol{x}_B(s))\,c_{13}(\boldsymbol{x}_B(s)) \\
&+ p_1(\boldsymbol{x}_B(s))\,I_3(\boldsymbol{x}_B(s)) - p_1(\boldsymbol{x}_B(s))\,q_1(\boldsymbol{x}_B(s))\,b(\boldsymbol{x}_B(s))]\,ds \\
= &-\iint_A \left[p_1(\boldsymbol{x})\,\mathrm{div}\,\varphi_1(\boldsymbol{x}) + \varphi_1(\boldsymbol{x})\,\mathrm{grad}\,p_1(\boldsymbol{x})\right] dx_1dx_2 \\
&+ \int_{\partial A} [p_1(\boldsymbol{x}_B(s))\,q_{12}(\boldsymbol{x}_B(s)) + p_1(\boldsymbol{x}_B(s))\,q_{13}(\boldsymbol{x}_B(s)) \\
&+ p_1(\boldsymbol{x}_B(s))\,q_{14}(\boldsymbol{x}_B(s)) + p_1(\boldsymbol{x}_B(s))\,c_{13}(\boldsymbol{x}_B(s)) \\
&+ p_1(\boldsymbol{x}_B(s))\,I_3(\boldsymbol{x}_B(s)) \\
&- p_1(\boldsymbol{x}_B(s))\,q_1(\boldsymbol{x}_B(s))\,b(\boldsymbol{x}_B(s))]\,ds.
\end{aligned}
\tag{5.92}
$$

Additionally, $\mathit{div}\, p_1(\boldsymbol{x})\, \varphi_1(\boldsymbol{x}) = p_1(\boldsymbol{x})\, \mathit{div}\, \varphi_1(\boldsymbol{x}) + \varphi_1(\boldsymbol{x})\, \mathit{grad}\, p_1(\boldsymbol{x})$ holds. Thus, Eq. (5.92) can further be transformed as:

$$
\begin{aligned}
\iint_A \pi_{Tq1}(\boldsymbol{x})\,dx_1dx_2 = &-\iint_A \mathrm{div}\, p_1(\boldsymbol{x})\,\varphi_1(\boldsymbol{x})\,dx_1dx_2 \\
&+ \int_{\partial A} [p_1(\boldsymbol{x}_B(s))\,q_{12}(\boldsymbol{x}_B(s)) + p_1(\boldsymbol{x}_B(s))\,q_{13}(\boldsymbol{x}_B(s)) \\
&+ p_1(\boldsymbol{x}_B(s))\,q_{14}(\boldsymbol{x}_B(s)) + p_1(\boldsymbol{x}_B(s))\,c_{13}(\boldsymbol{x}_B(s)) \\
&+ p_1(\boldsymbol{x}_B(s))\,I_3(\boldsymbol{x}_B(s)) \\
&- p_1(\boldsymbol{x}_B(s))\,q_1(\boldsymbol{x}_B(s))\,b(\boldsymbol{x}_B(s))] \\
= &-\int_{\partial A} [p_1(\boldsymbol{x}_B(s))\,q_{12}(\boldsymbol{x}_B(s)) + p_1(\boldsymbol{x}_B(s))\,q_{13}(\boldsymbol{x}_B(s)) \\
&+ p_1(\boldsymbol{x}_B(s))\,q_{14}(\boldsymbol{x}_B(s)) + p_1(\boldsymbol{x}_B(s))\,c_{13}(\boldsymbol{x}_B(s)) \\
&+ p_1(\boldsymbol{x}_B(s))\,I_3(\boldsymbol{x}_B(s)) \\
&- p_1(\boldsymbol{x}_B(s))\,q_1(\boldsymbol{x}_B(s))\,b(\boldsymbol{x}_B(s))]\,ds \\
&+ \int_{\partial A} [p_1(\boldsymbol{x}_B(s))\,q_{12}(\boldsymbol{x}_B(s)) + p_1(\boldsymbol{x}_B(s))\,q_{13}(\boldsymbol{x}_B(s)) \\
&+ p_1(\boldsymbol{x}_B(s))\,q_{14}(\boldsymbol{x}_B(s)) + p_1(\boldsymbol{x}_B(s))\,c_{13}(\boldsymbol{x}_B(s)) \\
&+ p_1(\boldsymbol{x}_B(s))\,I_3(\boldsymbol{x}_B(s)) \\
&- p_1(\boldsymbol{x}_B(s))\,q_1(\boldsymbol{x}_B(s))\,b(\boldsymbol{x}_B(s))]\,ds = 0.
\end{aligned}
\tag{5.93}
$$

Equation (5.93) asserts that the total profit of the C.T.A. in the city area becomes zero.

In turn, we formulate the behavior of the labor transport agents (L.T.A1. and L.T.A2.) (Beckmann 1952; Beckmann and Puu 1985; and Puu 2003). The L.T.Ai. receives wages of $w_i(\boldsymbol{x})l_{di}(\boldsymbol{x})b(\boldsymbol{x})$ from firms at location $\boldsymbol{x}$ and pays wages of $w_i(\boldsymbol{x})l_{si}h_i(\boldsymbol{x})$ to households of type i. As mentioned earlier, it is assumed that there is no migration from and to the outside of the city during equilibrium. Therefore, the profit in the L.T.Ai. at $\boldsymbol{x}$ is denoted as follows:

$$
\begin{aligned}
\pi_{Tli}(\boldsymbol{x}) &= w_i(\boldsymbol{x})\,ld_i(\boldsymbol{x})\,b(\boldsymbol{x}) - w_i(\boldsymbol{x})\,ls_i h_i(\boldsymbol{x}) \\
&= -w_i(\boldsymbol{x})\ \mathrm{div}\,\psi_i(\boldsymbol{x}) - w_i(\boldsymbol{x})\,b_i \left|\left|\psi_1(\boldsymbol{x})\right|\right|.
\end{aligned}
\tag{5.94}
$$

The L.T.Ai. determines labor transport routes to maximize the profit which gained over the entire city area. The optimal condition can also be obtained by applying the calculus of variation.

$$w_i(\boldsymbol{x})\, b_i \frac{\psi_i(\boldsymbol{x})}{\|\psi_i(\boldsymbol{x})\|} = \text{grad}\, w_i(\boldsymbol{x}). \tag{5.95}$$

5.13 Commodity Transport Agent (C.T.A.)

Transportation of commodities is assumed to be performed by the C.T.A. (Beckmann 1952; Beckmann and Puu 1985; and Puu 2003). The C.T.A. buys $p_2(\boldsymbol{x}_B)q_2(\boldsymbol{x}_B)$ of agricultural products at a point on the city boundary, $(\boldsymbol{x}_B)$, and sells $p_2(\boldsymbol{x})q_{21}(\boldsymbol{x})b(\boldsymbol{x}) + p_2(\boldsymbol{x})c_{21}(\boldsymbol{x})h_1(\boldsymbol{x}) + p_2(\boldsymbol{x})c_{22}(\boldsymbol{x})h_2(\boldsymbol{x})$ commodities to firms and households. Thus, the profit of C.T.A. at $\boldsymbol{x}$ is expressed as follows:

$$\begin{aligned}\pi_{Tq2}(\boldsymbol{x}) &= p_2(\boldsymbol{x})\, q_{21}(\boldsymbol{x})\, b(\boldsymbol{x}) + p_2(\boldsymbol{x})\, c_{21}(\mathbf{x})\, h_1(\boldsymbol{x}) \\ &\quad + p_2(\boldsymbol{x})\, c_{22}(\boldsymbol{x})\, h_2(\boldsymbol{x}) \\ &= -p_2(\boldsymbol{x})\,\text{div}\,\varphi_2(\boldsymbol{x}) - p_2(\boldsymbol{x})\, a_2\, \|\varphi_2(\boldsymbol{x})\|,\end{aligned} \tag{5.96}$$

$$\begin{aligned}\pi_{Tq2}(\boldsymbol{x}_B) &= p_2(\boldsymbol{x}_B)\, q_{21}(\boldsymbol{x}_B)\, b(\boldsymbol{x}_B) + p_2(\boldsymbol{x}_B)\, c_{21}(\boldsymbol{x}_B)\, h_1(\boldsymbol{x}_B) \\ &\quad + p_2(\boldsymbol{x}_B)\, c_{22}(\boldsymbol{x}_B)\, h_2(\boldsymbol{x}_B) - p_2(\boldsymbol{x}_B)\, q_{n2}(\boldsymbol{x}_B) \\ &= -p_2(\boldsymbol{x}_B)\,\text{div}\,\varphi_2(\boldsymbol{x}_B) - p_2(\boldsymbol{x}_B)\, a_2\, \|\varphi_2(\boldsymbol{x}_B)\|,\end{aligned} \tag{5.97}$$

$$\begin{aligned}\therefore \iint_A \pi_{Tq2}(\boldsymbol{x})\, dx_1 dx_2 &= -\iint_A \left[p_2(\boldsymbol{x})\,\text{div}\,\varphi_2(\boldsymbol{x}) + p_2(\boldsymbol{x})\, a_2\, \|\varphi_2(\boldsymbol{x})\|\right] dx_1 dx_2 \\ &\quad - \int_{\partial A} p_2(\boldsymbol{x}_B(s))\, q_{n2}(\boldsymbol{x}_B(s))\, ds, \\ &= -\iint_A \text{div}\, p_2(\boldsymbol{x})\, \varphi_2(\boldsymbol{x})\, dx_1 dx_2 - \int_{\partial A} p_2(\boldsymbol{x}_B(s))\, q_{n2}(\boldsymbol{x}_B(s))\, ds \\ &= \int_{\partial A} p_2(\boldsymbol{x}_B(s))\, q_{n2}(\boldsymbol{x}_B(s))\, ds - \int_{\partial A} p_2(\boldsymbol{x}_B(s))\, q_{n2}(\boldsymbol{x}_B(s))\, ds = 0.\end{aligned} \tag{5.98}$$

Similarly, we obtain the zero-profit equations:

$$\iint_A \pi_{Tqk}(\boldsymbol{x})\, dx_1 dx_2 = 0 \;\; (k = 3, 4). \tag{5.99}$$

5.14 Commodity Prices and Wage Rates

Equation (5.89) can be transformed into the following nonlinear first order partial differential equation:

$$\left(\frac{\partial \ln p_1(\boldsymbol{x})}{\partial x_1}\right)^2 + \left(\frac{\partial \ln p_1(\boldsymbol{x})}{\partial x_2}\right)^2 = a_1^2. \tag{5.100}$$

This equation can mathematically be solved. Here, we consider a special case where the initial manifold is degenerated to one point. Thus, the initial condition of $x_1(s,t)$, $x_2(s,t)$ and $\ln p_1(x_1,x_2)$ is specified as $x_1(0,t) = 0$, $x_2(0,t) = 0$, and $\ln p_1(0,t) = p_{10}$, respectively. Then the solution surface for the commodity price is written as Eq. (5.101). This equation represents a cone with a vertex of $(0, 0, p_{10})$ and is said to be integral conoid (Courant and Hilbert 1953, 1962).

$$p_1(x_1, x_2) = p_{10} \exp\left(a_1\sqrt{x_1^2 + x_2^2}\right). \tag{5.101}$$

Similarly, solution surfaces of other prices are as follows:

$$p_k(x_1, x_2) = p_{k0} \exp\left(-a_k\sqrt{x_1^2 + x_2^2}\right) \ (k = 2, 3, 4). \tag{5.102}$$

Regarding wage rates, the same discussion is possible. The solution surface for the wage rate, *i*, is expressed as:

$$w_k(x_1, x_2) = w_{k0} \exp\left(-b_k\sqrt{x_1^2 + x_2^2}\right) \ (k = 1, 2). \tag{5.103}$$

Assuming the business district encircles the origin, and the residential area is located outside the business area (i.e., a simple von Thünen ring), the wage rate prevailing at the city center, w_{k0}, is given by the wage rate at the city boundary, which is determined to satisfy the equilibrium conditions mentioned below. Contrary to the urban commodity price surface, the wage profile shows a decrease of exponential order from the city center to the city boundary.

5.15 Equilibrium Conditions

Skipping other details because of page limitations, we can finally derive the following equilibrium conditions.

Urban Good Market

$$\begin{aligned} \left\|\varphi_1(\boldsymbol{x}(S))\right\| = \int_0^S \Big[& q_q(\boldsymbol{x}(\tau))\, b(\boldsymbol{x}(\tau)) - c_{11}(\boldsymbol{x}(\tau))\, h_1(\boldsymbol{x}(\tau)) \\ & - c_{12}(\boldsymbol{x}(\tau))\, h_2(\boldsymbol{x}(\tau)) \\ & - I_1(\boldsymbol{x}(\tau))\, b_2(\boldsymbol{x}(\tau))\Big] \left(\tfrac{\tau}{S}\right) \exp\ a_1(\tau - S)\, d\tau \\ = & \sum_{j=2}^{4} q_{n1j} + N_3 c_{n13} + I_{n3}. \end{aligned} \tag{5.104}$$

Agricultural, Forestry and Rural Goods Market

$$\begin{aligned} \|\varphi_k(\boldsymbol{x}(0))\| &= \lim_{s\to S}\int_0^S [-q_{k1}(\boldsymbol{x}(S-\tau))\,b(\boldsymbol{x}(S-\tau)) \\ &\quad - c_{k1}(\boldsymbol{x}(S-\tau))\,h_1(\boldsymbol{x}(S-\tau)) \\ &\quad -c_{k2}(\boldsymbol{x}(S-\tau))\,h_2(\boldsymbol{x}(S-\tau))]\left(\tfrac{S-\tau}{S-s}\right)\exp\ a_k(\tau-S)\,d\tau \\ &\quad + q_{nk} = 0\ (k=2,3,4), \end{aligned} \tag{5.105}$$

$$q_k = \int_{\partial A} q_{nk}(S)dS + \sum_{j=2}^{4} q_{kj} + c_{k3} \quad (k=2,3,4). \tag{5.106}$$

Labor Market for H.I.H.s

$$\begin{aligned} \left\|\psi_1(\boldsymbol{x}(0))\right\| &= \lim_{s\to S}\int_0^S [l_{s1}h_1(\boldsymbol{x}(S-\tau)) \\ &\quad -l_{d1}(\boldsymbol{x}(S-\tau))\,b(\boldsymbol{x}(S-\tau))]\left(\tfrac{S-\tau}{S-s}\right)\exp\ b_1(\tau-S)\,d\tau = 0. \end{aligned} \tag{5.107}$$

Labor Market for L.I.H.s

$$\begin{aligned} \left\|\psi_2(\boldsymbol{x}(0))\right\| &= \lim_{s\to S}\int_0^S [l_{s2}h_2(\boldsymbol{x}(S-\tau)) \\ &\quad -l_{d2}(\boldsymbol{x}(S-\tau))\,b(\boldsymbol{x}(S-\tau))]\left(\tfrac{S-\tau}{S-s}\right)\exp\ b_2(\tau-S)\,d\tau = 0. \end{aligned} \tag{5.108}$$

Labor Market in the Rural Area

$$N_3 = ld_2 + ld_3 + ld_4. \tag{5.109}$$

Capital Market in the Urban Area

$$KS_1 = \iint_A kd_1(\boldsymbol{x})\,b(\boldsymbol{x})\,dx_1dx_2. \tag{5.110}$$

Capital Market in the Rural Area

$$KS_3 = kd_2 + kd_3 + kd_4. \tag{5.111}$$

Land Market
See Eqs. (5.66), (5.67), (5.68), (5.69), (5.70), (5.71), (5.72) and (5.73).

Location Equilibrium Conditions

$$\pi_{B1}(\boldsymbol{x}) = 0 \ \ (\boldsymbol{x} \in \text{ business area}). \tag{5.112}$$

$$u_1(\boldsymbol{x}) = u_1^* \ \ (\boldsymbol{x} \in \text{ normal land}). \tag{5.113}$$

$$u_2(\boldsymbol{x}) = H_3^* \ \ (\boldsymbol{x} \in \text{ flood} - \text{prone area}). \tag{5.114}$$

H_3^* is the rural utility level, endogenously determined.

Constraints on the Numbers of Firms and Households

$$M = \iint_{A_B} \frac{1}{m_B(\boldsymbol{x})} dx_1 dx_2. \tag{5.115}$$

$$N_1 = \iint_{A_{H1}} \frac{1}{m_{H1}(\boldsymbol{x})} dx_1 dx_2. \tag{5.116}$$

$$N_2 = \iint_{A_{H2}} \frac{1}{m_{H2}(\boldsymbol{x})} dx_1 dx_2. \tag{5.117}$$

$$N_3 : \text{exogenously given}. \tag{5.118}$$

In these equations, S, A_B, A_{H1}, and A_{H2} depict the parameter value expressing the city boundary, business area, normal land, and flood-prone area, respectively. The model describes the rural–urban–natural environment–flood interaction observed in Palangkaraya City in Indonesia.

5.16 Comparative Static Analysis

Here, we consider the stationary state, and we derive some propositions from the comparative static analyses.

Proposition 1

The flood-prone areas are occupied by L.I.H.s, whereas the normal land is occupied by H.I.H.s. The reason is that $\partial g_{Hi}/\partial Y_i < 0$ holds from Eq. (5.27). Therefore, if the per capita income increases, then the bid-rent function decreases. Thus, the flood-prone areas are occupied by L.I.H.s.

Proposition 2
A slight increase in income of L.I.H. decreases the bid-rent and increases the bid-max lot size. Thus, parameter θ_i in Eq. (5.23) can decrease the number of L.I.H. in the flood-prone areas.

Proposition 3
An increase in the residential area of the rural region increases the current value of the Hamiltonian. Thus, the supreme utility of L.I.H. in the flood-prone areas is increased. Thus, the number of L.I.H. is decreased in the flood-prone areas.

Proposition 4
If the carrying capacity of forests in the rural area is increased, the bid-rent by L.I.H. is increased, and the bid-max lot size is decreased, resulting in an increase in the population in the flood-prone areas, although flood risk is decreased.

Proposition 5
If the von Thünen parameters, $a_k(k = 1, 2, 3, 4)$, $b_i(i = 1, 2)$, are decreased by a transportation project, the business district and residential area (i.e., normal land) increase, and the utility level of H.I.H. increases. In the flood-prone areas, the bid-max lot size decreases, leading to an increase in the number of L.I.H.

5.17 Concluding Remarks

The authors have developed partial and general equilibrium urban economic models for Palangkaraya City. Through these models, we can conclude that the bid-rent of the L.I.H. is higher than that of the H.I.H. in flood-prone areas. This is caused by the introduction of the expected damage on household assets. This result is contrary to the result of traditional urban economics.

Flooding is of great concern in Palangkaraya City, and one of its causes is the harvesting of forests in rural areas surrounding the city. Hence, one must consider the socioeconomic activities and forests in the rural areas in parallel to the environmental impact. This point is a key motivation of this study. Moreover, it is important to consider the plane city rather than a linear city for reality.

In this study, Palangkaraya City was regarded as a plane city, and the external area of the city was assumed to be a point area (i.e., dimensionless). In the city and the external areas, the natural environmental level and the externality of forests are considered. The role of forests is to reduce flood damage to the city. Additionally, forests are able to improve the natural environmental levels in the city as well as in the external areas.

The important policy target of the Palangkaraya City government is the reduction of illegal settlements in flood-prone areas. In order to evaluate the policy, a comparative static analysis was done between it and an increase in the supreme utility level in the external area, a redistribution income policy, and fostering the forest in the external area. This study suggests the possibility of a reduction in illegal settlements in the flood-prone areas.

As a result of the fact that this analysis depends heavily on specific parameter values, it was necessary to estimate parameters by employing empirical data and presenting more realistic policies. These are left for future studies.

References

Alonso W (1964) Location and land use. Harvard University Press, Cambridge, MA
Anas A, Arnott R, Small K (1998) Urban spatial structure. J Econ Lit 36(3):1,426–1,464
Beckmann M (1952) A continuous model of transportation. Econometrica 20:642–660
Beckmann M, Puu T (1985) Spatial economics: density, potential, and flow. North-Holland, Amsterdam
Black D, Henderson V (1999) A theory of urban growth. J Polit Econ 107(2):252–284
Courant R, Hilbert D (1953) Methods of mathematical physics Vol.1. Interscience Publishers, New York/London
Courant R, Hilbert D (1962) Methods of mathematical physics Vol.2. Interscience Publishers, New York
Fujita M (1989) Urban economic theory: land use and city size. Cambridge University Press, Cambridge
Fujita M, Ogawa H (1982) Multiple equilibria and structural transition of non-monocentric urban configurations. Reg Sci Urban Econ 12:161–196
Fujita M, Krugman P, Venables A (1999) The spatial economy: cities, regions, and international trade. The MIT Press, Cambridge, MA
Gelfand IM, Fomin SV (1963) The problem in calculus variation (trans: Silverman R). Prentice–Hall
Higano Y, Shibusawa H (1999) Agglomeration diseconomies of traffic congestion and agglomeration economies of interaction in the information-oriented city. J Reg Sci 39(1):21–49
Hörmander L (1990a) The analysis of linear partial differential operators I. Springer, Berlin
Hörmander L (1990b) The analysis of linear partial differential operators IV. Springer, Berlin
Leeuwen ES, Nijkamp P (2006) The urban–rural nexus – a study on extended urbanization and the hinterland. Stud Reg Sci 36(2):283–303
Lucas R, Rossi–Hansberg E (2002) On the internal structure of cities. Econometrica 70(4):1,445–1,476
Mills ES (1967) An aggregative model of resource allocation in a metropolitan area. Am Econ Rev 57:197–210
Miyata Y (2011) Integrating commodity and labor flows into monocentric city over a two dimensional continuous space. Stud Reg Sci 39(3):631–658
Miyata Y, Shibusawa H, Permana I, Wahyuni A (2018) Environmental and natural disaster resilience of Indonesia. Springer, Singapore
Muth RF (1969) Cities and housing. University of Chicago Press, Chicago/London
Permana I, Miyata Y (2009) Analysis of illegal settlements in flood prone areas in Central Kalimantan Province, Indonesia -a two dimensional spatial modeling. In: Proceedings of the 46th annual meetings of Japan Section Regional Science Association International, CD-ROM
Permana I, Miyata Y (2012a) An urban economic model of illegal settlements in flood prone areas in Palangkaraya city, Indonesia – a partial equilibrium analysis. Reg Sci Inq 4(1):29–38
Permana I, Miyata Y (2012b) Analysis of illegal settlements in flood prone areas in Parangkaraya city, Indonesia – a general equilibrium modeling approach. J Environ Hum Symbiosis 19:65–78
Puu T (2003) Mathematical location and land use theory. Springer, Berlin
Richardson HW (1977) The new urban economics: and alternatives. Pion, London

Shibusawa H (2000) Cyberspace and physical space in an urban economy. Pap Reg Sci 79(3):253–270
Solow RM (1973) On equilibrium models of urban locations. In: Parkin JM (ed) Essays in modern economics. Longman, London, pp 2–16
von Thünen JH (1826) Der isolirte Staat in Beziehung auf Landwirthschaft und Nationalökonomie. Hamburg

Chapter 6
Environmental Assessment of Biomass Energy Crops

Susumu Uchida

Abstract Biomass energy has an advantage among renewable energy technologies because it is storable potential energy. A variety of biomass energy technologies have been subjects of research on their environmental burdens. Because these technologies are strongly related to agricultural production, their evaluation has faced the issue of uncertainty that is inherent in agriculture. Methodologies in assessing the environmental impact of biomass energy technologies, including how that uncertainty should be treated in the assessment, are introduced in this chapter. This is followed by some case studies, including those of the life cycle assessment of energy crop cultivation and evaluation of the effect of economic promotion policies on greenhouse gas reduction.

Keywords Biomass · Energy crop cultivation · Life cycle assessment · Promotion policy · Model simulation

6.1 Biomass Energy Technologies

Renewable energy technologies are being developed globally as a countermeasure to environmental issues related to climate change and the depletion of fossil fuel resources. Low-cost solar and wind power generation technologies have become widespread. However, the corresponding energies—radiation energy and kinetic energy—cannot be stored independently since they are flow-type energies. In contrast, biomass energy production based on chemical energy is a well-known technology, which produces a storable potential energy (Whitaker et al. 2010). Biomass energy technologies are considered to include first- to third-generation technologies (Ho et al. 2014). Technologies based on energies from fuel crop production are considered to be first-generation technologies. These have been

---

S. Uchida (✉)
Department of Agriculture, Ibaraki University, Mito, Japan
e-mail: uchida@mx.ibaraki.ac.jp

K. Nakayama, Y. Miyata (eds.), *Theoretical and Empirical Analysis in Environmental Economics*, New Frontiers in Regional Science: Asian Perspectives 34, https://doi.org/10.1007/978-981-13-2363-8_6

developed since the early years of energy-based technology development, and examples of their application include bioethanol production from corn in the United States and from sugarcane in Brazil. However, these first-generation biomass technologies have raised some controversies, because of the conflict between the production of these crops for fuel and food production. To solve this problem, fuel technologies using nonfood biomass, which include rice straw and woods, have been examined and are considered to be second-generation technologies. Although producing energy from these sources does not compete with food production, energy conversion of these sources is difficult since their energy densities are relatively low. Therefore, their practical usage has been restricted thus far. The two generations above both rely on conversion technology that generates alcohol from starch or sugar. However, third-generation technology is quite different in that oil is extracted from algae. This technology is expected to flourish in the future because its energy efficiency is high, and no energy conversion is needed. Oil is generated directly from algae, and it is available only by squeezing. Second- and third-generation technologies are under development, while first-generation technology currently utilizes the most biomass energy available for practical use.

First-generation biofuel crops are crops that are generally used for food production; however, the functionality required is different from that of food crops. Although both crop types are energy sources for humans, food supplies energy required by the body to sustain life, while biofuel provides energy for higher-order economic and civil activities. Furthermore, in the case of food, there are various secondary requirements, such as flavor, absorbability, and the absence of harmful contaminants, while in the case of biofuels, the only requirement, in principle, is high energy density. Considering these differences, crop varieties with higher energy density need to be especially developed for biofuel production. When introducing the use of biomass energy as a countermeasure to environmental issues, it is important to consider the degree to which environmental considerations take priority over conventional energy technologies in decision-making. Methods for evaluating this degree include the life cycle assessment (LCA), environmental impact assessment (EIA) (Bradley 1975), and strategic environmental assessment (SEA) (Therivel 1993). Since LCA is the most versatile and commonly applied method, it will be this chapter's main focus, with a detailed biomass energy evaluation in relevant case studies.

6.2 Evaluation of Biomass Energy

An LCA is usually conducted based on a dataset, known as an inventory database (Uchida et al. 2010a). LCA databases, such as ecoinvent and IVAM LCA Data, have been constructed for various countries. In Japan, the Inventory Database for Environmental Analysis (IDEA), developed by the National Institute of Advanced Industrial Science and Technology and the Japan Environmental Management Association for Industry, covers general products, apart from certain services. However, evaluating biomass energy using these general databases alone is presently difficult, primarily because of the absence of an established industry. As a result, many issues regarding the availability and reliability of data persist. Another important reason is the difficulty faced in creating databases for the agricultural production sector, which is an important element in the evaluation process. For biomass energy fueled by energy crops, the LCA of the cultivation process considers three notable variable factors. The first factor is the diversity in agricultural production processes and the corresponding use of material inputs. The timing of processes, such as sowing and tilling, and the use of fertilizers, agrichemicals, and soil improvers depend significantly on the conditions of the weather and soil. Therefore, they vary from region to region and, ultimately, from farm to farm. The same producer may witness different results from year to year, depending on the weather in a given year, the conditions in the previous year (resulting from the kind of cultivation practiced on that land in the previous year), and other factors. Therefore, establishing representative values for agricultural production is a difficult task. The LCA results for biomass energy also cover a wide range from positive to negative values. These results probably reflect the uncertainty due to the abovementioned diversity. The second factor is the uncertainty in the evaluation of the volume of gas emissions from the soil. The emissions of CO_2, CH_4, and N_2O, all of which have a significant global warming potential, are likely to have a considerable impact on the overall results of rice paddies and fields. However, the volumes vary from case to case and with the abovementioned factors. Another uncertainty arises from the fact that biomass energy production is likely to involve the conversion of nonagricultural land as well as conventional farming production. Therefore, the amount of gas emitted depends heavily on the condition of the source land. The third factor comprises issues regarding the scope of LCA evaluation. Most of the research examples related to the evaluation of biomass energy have a scope limited to global warming and energy supply balance. However, for certain domains of environmental impact (related to global warming and nutrition), the environmental load of biomass energy may exceed that of fossil fuels. Therefore, there is a need for a domain-specific comprehensive evaluation of such research examples for various domains.

To respond to these uncertainties, the inventory data values are determined through a combination of approaches. In the first approach, the actual measured data from the cultivated land is used, as it is the most accurate data available. This may include site surveys by farmers and experimental data generated through the testing of cultivated land by research organizations. Next, to obtain general data for each region, standard technology systems from public institutions may be used. This mainly comprises standard operations for each crop formulated by the respective municipalities. This provides for a certain level of representation of regional diversity. However, it should be noted that this data does not necessarily reflect the actual status of the region. In cases where individual data cannot be obtained, it is necessary to estimate the data values using statistical data, such as management data (agriculture and forestry census), expense data (production expense surveys), and industrial data related to individual materials, such as agricultural machinery and fertilizers.

For data that cannot be obtained, even by statistical means, simulation modeling is effective. For example, the emission volumes of greenhouse gases from soil, which involve significantly uncertain factors, can be estimated using dynamic models for nitrogen and carbon. Since the main greenhouse gases, CO_2, CH_4, and N_2O, are generated from the reaction of carbon and nitrogen in soil, ascertaining the dynamic states of these elements enables the determination of the respective emission volumes. One of the models used for analyzing the dynamic states of both nitrogen and carbon is the DNDC model (Li et al. 1992). Meanwhile, various models have been developed for analyzing the dynamic states of only nitrogen (an example is SOILN_jpn, the Japanese adaptation of the SOILN model) (Maeda 2008). To analyze the dynamic state of carbon, the RothC model is widely used; its modified version has been developed for Japan, with an advanced applicability to rice paddies (Shirato et al. 2004). To model the dynamic state of carbon in a forest, the globally used CENTURY model has been fine-tuned for Japan to develop SENTURY-jfos. All these models use weather conditions, soil status, and cultivated crop (vegetation in the case of forests) as inputs and are used to estimate the emission volume or underground eluviation amount of a target substance. Since they analyze the dynamic state comprehensively, without being limited to greenhouse gases, it is also possible to evaluate other domains of environmental impact. This resolves the evaluation scope issue mentioned earlier in the third factor. The CropWat model has been recommended for the analysis of issues related to water resource consumption, a potentially important global environmental issue in the future. This model can provide estimates from similar inputs, as mentioned above. An input-output table (I/O table) is utilized for the estimation of background data, including that regarding the usage amounts of materials when these amounts are difficult to obtain. This table covers all industries and presents monetary data that can be used to derive the usage data. However, the monetary data is only summarized by industry. Since specific data of the average prices of each material is required, when such data are unavailable, accurate usage results are not obtained.

6.3 Examples of Assessment of Bioethanol

There have been many previous studies on the environmental impact assessment of biomass energy; however, most of these studies have focused on the energy balance and global warming potential of a single fuel crop. This chapter first discusses the assessment of various environmental impact categories and then covers environmental load reduction through the reciprocal use of local resources. Finally, this chapter examines how political measures, such as environmental restrictions and taxes, affect economic changes depending on the presence or absence of biomass energy technologies. This last example focuses on waste-type biomass rather than crop-type biomass.

6.3.1 Assessment of Various Environmental Impact Categories

Depending on the type of raw material, there have been many previous studies assessing the environmental impact of biomass energy, but there are few examples that have addressed categories other than energy balance and global warming potential. The authors of this chapter estimated the environmental load in producing bioethanol, from raw ingredients to shipment, with regard to five types of energy crop and eight environmental impact categories (Uchida and Hayashi 2012). The assessments here were conducted using data that reflect the results of expected improvements in cultivation techniques and cultivars, in order to investigate the extent of the potential of so-called "first-generation" crop-type bioethanol in reducing the environmental load. Figure 6.1 shows a portion of the assessment results. Energy resource consumption in (b) is found on an energy basis, so it represents the consumption of energy derived from fossil fuels. The results are expressed relative to gasoline, and the graphs show that the superiority of bioethanol varies greatly depending on the category. The ratio of output from cultivation to output from energy conversion also depends heavily on the category.

6.3.2 Assessment of System that Includes Reciprocal Use of Biomass Resources

Biomass energy relies heavily on the energy resources existing in the locality, so various ripple effects in the local area can be expected. Particularly in terms of the environment, overall load could be reduced through the reciprocal use of local resources. Figure 6.2 shows an example of this kind of model (Uchida et al. 2010b). The scenario in this example assumes the introduction of bioethanol production into an area in the Minami-Kyushu region of Japan. The heavy-lined items are newly introduced industries, and bioethanol is produced by cultivating sweet potatoes on abandoned agricultural land and using them as a raw material. A feature of this model is the reciprocal use of resources by another major industry in the area, the livestock industry. The stems and leaves that grow during cultivation and the residue from fermentation are mixed into the feed, together with the fermentation residue from starch and shochu (an alcoholic beverage distilled from sweet potatoes), and this is used in the livestock industry. Also, manure produced by the livestock industry is composted and used in sweet potato cultivation. Compost is currently transported to other regions because there is an excess. However, the environmental load associated with this transportation can be reduced by establishing this kind of cycle within the local area.

There are several methods of evaluating this kind of cyclical model in LCA. Broadly speaking, there are methods that divide the system and methods that extend the system and evaluate it as a whole. When the aim is to evaluate the entire system, as in this case, a system extension method is used. However, typical LCA tools connect modules in one direction (raw ingredients → production → consumption → disposal), so they do not allow the creation of process loops. In that case,

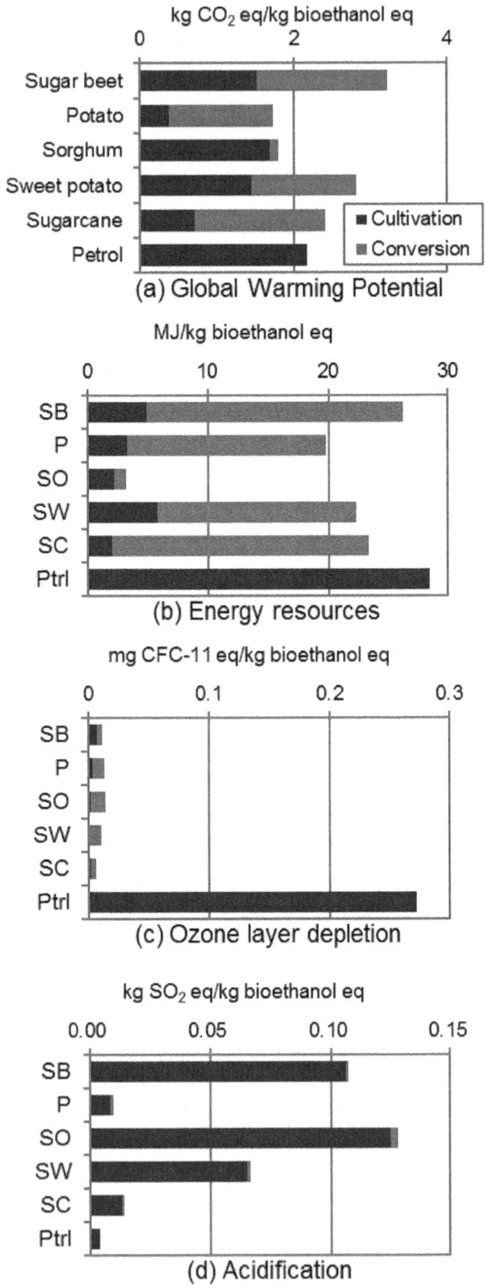

Fig. 6.1 Environmental impacts associated with bioethanol production. (**a**) Global warming potential. (**b**) Energy resources. (**c**) Ozone layer depletion. (**d**) Acidification

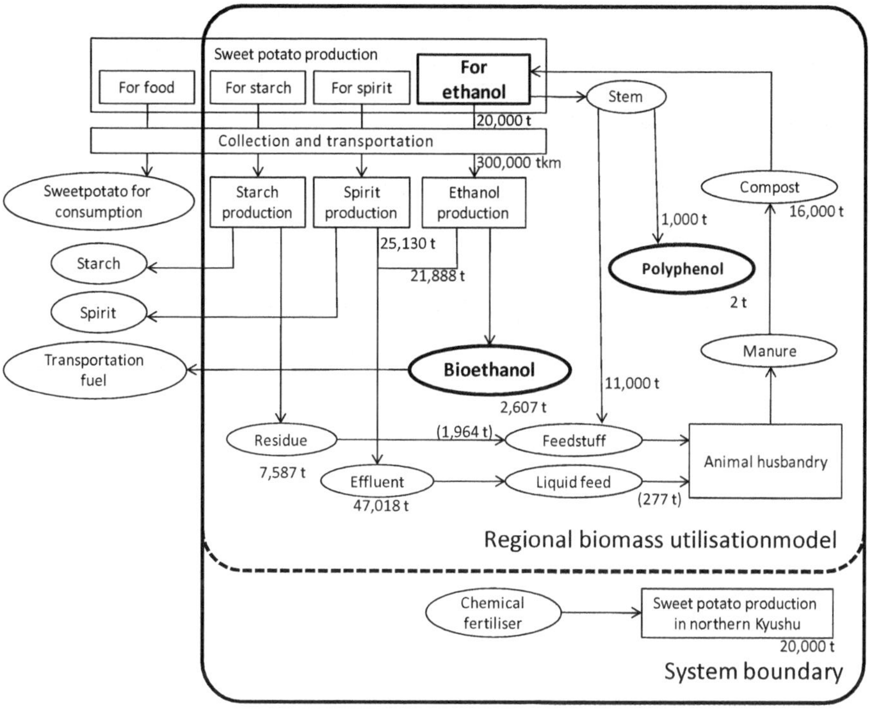

Fig. 6.2 Regional biomass utilization scenario

after correctly comprehending the material flows, it is necessary to disassemble the process flows and arrange them into linear flows before performing the analysis. This method does not present any problems in terms of evaluation, but because all flows must be clarified at the start, it is not suitable in cases where the material flows will be analyzed during the assessment process.

Figure 6.3 shows the results of such an environmental impact assessment of the model of biomass utilization through the reciprocal use of resources and a case in which biomass utilization is combined with improvements in cultivation techniques and cultivars mentioned in the preceding section, relative to a conventional model (including production of gasoline with the same energy content) (Uchida et al. 2010b). The figure shows eight environmental impact categories, and in the case in which technique/cultivar improvement is carried out in addition to biomass utilization, the categories, human toxicity (cancer) and terrestrial ecotoxicity, show an increase in environmental load compared to the conventional model. The load is the same or lower for all other categories. This is the result of the efficient use of biomass resources through collaboration with other industries at multiple stages. Thus, the efficient use of local resources takes full advantage of the regional characteristics of biomass energy and can be expected to show an enhanced environmental load reduction effect.

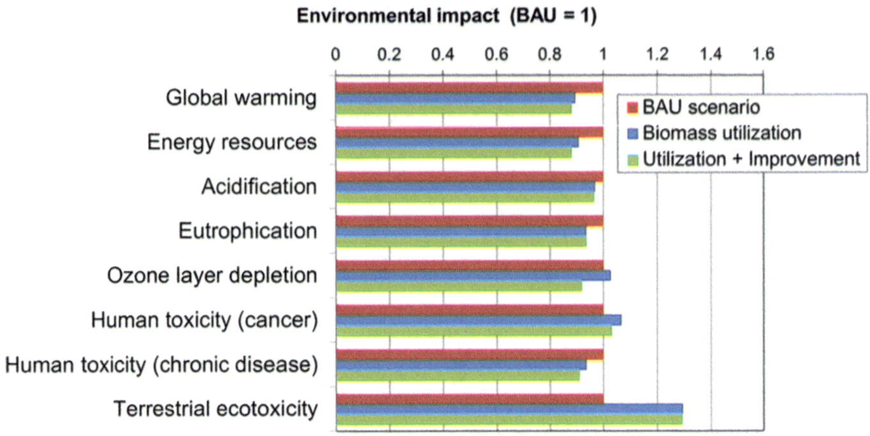

Fig. 6.3 Evaluation of the biomass utilization scenario

In cases such as this model that incorporate various improvements, it is difficult to determine the degree to which individual improvements contributed to the overall result. Therefore, regarding the individual items of (1) the introduction of bioethanol production, (2) use of fermentation residue, (3) use of stems and leaves, (4) direct usage of compost, and (5) improvement of techniques/cultivars, the authors constructed four hypothetical scenarios in which the number of items introduced was increased one by one and estimated the effect of each item by finding and comparing the respective environmental loads. Figure 6.4 shows the estimation results (Uchida et al. 2010b). The horizontal axis shows the environmental load reduction ratio relative to the conventional model, with the positive direction representing a reduction in load. Figure 6.4 shows that each item contributes to the reduction in environmental load to some extent. The results for ozone layer depletion and toxicities, in particular, demonstrate the advantage of combined improvements, as an increase in environmental load due to the introduction of bioethanol is offset by the other factors.

Since competition with food production began to be regarded as a problem, initiatives concerning crop-type biomass, such as using raw materials that are substandard and, thus, unsuitable as food, and producing crops on abandoned agricultural land, have become mainstream. This increases the implications of recycling, similar to the production of biogas and biodiesel from recycled livestock and cooking oil waste, respectively. It is quite likely that implementing these kinds of initiatives in a region will lead to the exploration of other recyclable resources in the region. In many regions where biomass energy has actually been introduced, efforts are being made toward the reciprocal use of various resources with other industries, as in this utilization model. Constructing a system model to assess the system as a whole would make more appropriate environmental impact assessments possible in these regions.

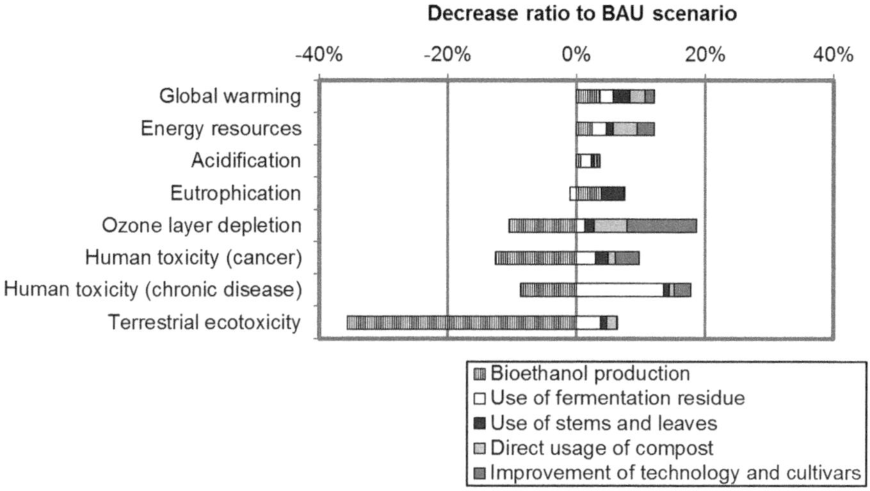

Fig. 6.4 Reduction factors of environmental impacts in the biomass utilization scenario

6.3.3 The Impact of Biomass Energy Use Under Environmental Policies on the Relationship Between Environment and the Economy

Environmental policies include regulatory measures and economic instruments, and easy-to-implement regulatory measures are commonly used. However, when environmental restrictions are applied, economic activity is simultaneously restricted. This problem is alleviated by substitution with low-environmental load technological advances, but without help, it is difficult for technologies requiring initial investment to become established. Here, we use a simulation based on an environment-economy model to examine the relationship between the economy and the environment with regard to a policy of simultaneously imposing environmental restrictions and an environmental tax (greenhouse gas tax), using the revenues to subsidize new technologies (Uchida et al. 2008). The environmental impact considered is greenhouse gas emissions, and the new technologies to be implemented are energy technologies that recycle waste (methane fermentation, synthetic gas, woody pellet, ethyl alcohol, dimethyl ether, biodiesel, waste power generation, compound waste power generation, fuel cell, waste oil, animal waste, and carcass bulky waste). First, total GDP over 11 years was determined for the imposition of three types of environmental restrictions in two scenarios—one scenario in which only the environmental restriction is imposed without technology substitution and the other scenario in which policies to promote the widespread use of new technologies through environmental taxes and subsidies are introduced. The three types of environmental restrictions are (in increasing order of strictness)

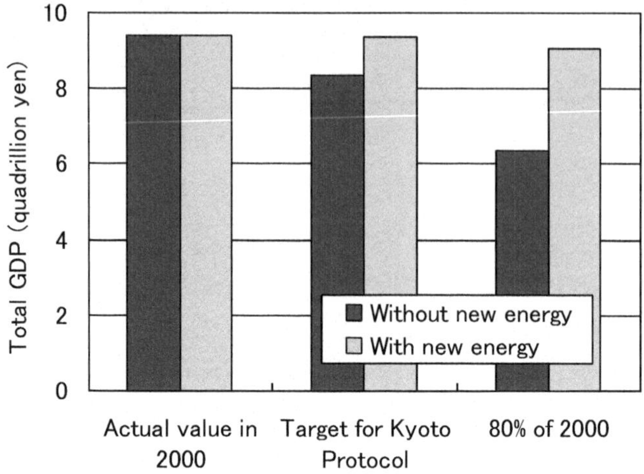

Fig. 6.5 Total GDP under various restrictions in GHG emissions

total greenhouse gas emissions from industries in Japan: at the 2000 level, at the Kyoto Protocol level, and at 80% of the 2000 level. Figure 6.5 shows the results. In the scenario without policies to promote widespread use of new technologies, a negative effect of environmental regulation on the economy is evident since GDP decreases as the total emission restrictions become stricter (Uchida et al. 2008). In contrast, little decrease in GDP is seen in the scenario where promotion policies are introduced, and the reduction in environmental load is achieved without negatively affecting the economy. The use of waste biomass energy can also be expected to have the direct effect of reducing the amount of waste discharged, and Fig. 6.6 shows the results of estimating that amount, under the policies to promote widespread use of new technologies mentioned above (Uchida et al. 2008). Waste is reduced by approximately 13% overall, with a large proportion of this reduction being in food waste. The largest direct reduction is in livestock waste, but the amount of sludge, which is a residue of using livestock waste to produce energy, increases, and when this increase is offset against the reduction, the quantitative reduction is small. Figure 6.7 shows the energy production from new technologies per unit GDP at different environmental tax rates (Uchida et al. 2008). The simulation results show that the quantity of biomass energy introduced increases linearly with the tax rate.

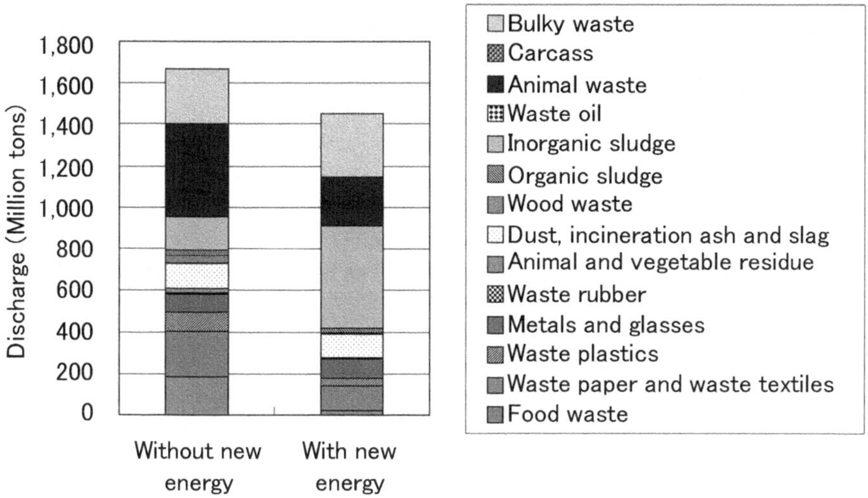

Fig. 6.6 Changes in total waste discharged due to introduction of new energy industries

Fig. 6.7 Dependency on new energy industries

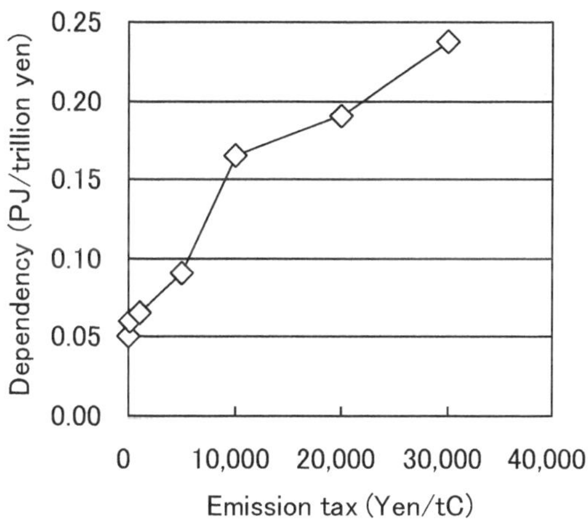

6.4 Future Direction of Assessment

6.4.1 Integration with Economic and Social Assessments

The effects of introducing biomass energy go beyond the environment. The fact that the production of biomass energy is driven by local activity means that its economic effects are assimilated within the region, including employment opportunities that previously left the region. Also, there are numerous examples of research on the social effects of biomass energy production, such as education about energy and environmental issues, and the effect of educating local residents about biomass energy is expected to accelerate the shift to a more sustainable society (Morimoto and Miyamoto 2009). Evaluating these kinds of economic and social impacts simultaneously with environmental impacts would produce a holistic assessment for the first time.

Currently, impacts in the three areas of environment, economy, and society are almost always assessed individually. However, hopes for assessments that integrate these areas have been growing for some time. Integration is possible in an LCA framework, but it involves applying weighting coefficients, and the method of deriving these coefficients and the reliability and validity of the coefficients obtained remain highly controversial. This problem is to a large extent dependent on the progress of future research.

Evaluation of interactions is also an issue in the integration of biomass energy with the economy and society. The introduction of biomass energy changes the economic and social conditions of a region, including its industrial structure. In principle, conventional LCA sums all the environmental loads emitted directly or indirectly in the past by an object already in existence (excluding use/disposal processes). In other words, it is a procedure that attributes various environmental loads that have already been emitted on earth to individual products. For this reason, conventional LCA is also called attributional LCA. In contrast, a method called consequential LCA includes the results of economic and social changes expected due to the manufacturing of the product in question, and examples of research using this method have increased sharply in recent years. (This reflects a demand for more realistic environmental impact assessments). Estimation of economic and social changes in consequential LCA is mainly performed using scenario analysis (Rehl et al. 2012). This scenario analysis is carried out by ascertaining the causal relationships and then incorporating changes into the scenario. This corresponds to the summation method in conventional LCA. The biomass utilization model introduced in Chap. 3 is also a type of consequential LCA, but because it focuses on the effects in the region, the causal relationships included in the object of the assessment are limited. There is also the method of estimating changes using a method corresponding to the input-output approach. When limited to the economy, it is considered possible to analyze the environmental impact, including economic ripple effects for each industry classification, using models based on economic theory such as input-output tables and general equilibrium theory, though there are few examples of research in this area, perhaps due to the problem of applying macro data to a limited area. Thus, a hybrid approach using the summation method and macro data is considered valid in consequential LCA. With regard to social change, it appears that more time is needed for the development of assessment methods because society is more strongly influenced by regional characteristics than is the economy. However, a possible direction is to first determine the relationships between production activities and society by using several factors such as employment, population, and place of residence as a foothold and then construct a database by summation and conversion to coefficients.

6.4.2 Response to Resource Problems

Another topic that will become increasingly important in environmental impact assessments is resource problems. Resources related to biomass, i.e., agriculture, include water and phosphorus resources.

There are concerns that water resources will become a serious problem over the course of this century. In addition to the increasing demand for water along with agricultural production, abnormal weather is causing local water shortages and flood damage. The problem could be exacerbated by changes in the rainfall distribution due to climate change. Here, we can point to three major differences between water resources and other resources, such as metals and fossil fuels, that are important when assessing water resource consumption. The first is that it is difficult for humans to control the supply rate. The supply rate of other resources, as industrial activities, can be changed to some extent, whereas water resources are generally supplied by rainfall and subsequent river water, and although they are stored in part, the long-term supply rate is dependent on rainfall. The second difference is that transportation is economically difficult. Apart from drinking water, which has high added value, the price of water is low relative to its weight, and its transportation cost is relatively high. In other words, it is difficult to transport water resources within an economic system in the same way as other resources. Consequently, although water resources are often transported by means of public works, the distances and volumes are limited, and water resource surpluses/deficits arise because it is difficult to mitigate regional variability. The third difference is that the reuse cycle is fast. The majority of fresh water on earth moves through a cycle of precipitation → flow to ocean → evaporation. Therefore, the problem of water resource supply depends heavily, not on the total amount of water but on the rate of supply to a certain area and at a certain point in time (Oki and Kanae 2007).

The above peculiarities indicate that a fundamental of water as a resource is the rate of supply rather than volume. The scarcity of other resources is expressed as the usage relative to availability or, in other words, as a stock concept. In contrast, in the case of water, it is more appropriate to express scarcity as the speed of use relative to the available flow rate or, in other words, as a flow concept.

Water footprint (WF) has been widely used as an indicator of water resource consumption based on the flow concept, and, in recent years, various indicators that also consider regional differences in availability and seasonal variation have been examined (Vanham and Bidoglio 2013; Ridoutt et al. 2012; Federal Office for the Environment FOEN 2009; Pfister et al. 2009). There are also examples of research that has developed indicators similar to the original footprint concept by conversion into area to be supplied with water (Stoeglehner et al. 2011; Gleeson et al. 2012). In these research examples, water resource consumption is evaluated by determining the flow of water consumption for an artificial unit of time such as 1 year or 1 month; however, considering the abovementioned peculiarities of water resources, a more fundamental flow of water use must be defined to accurately represent the consumption load relative to the flow of water supply, which varies according to region and time. Also, the development of an indicator of water resource consumption based on this kind of concept and its use as an environmental impact category will improve the accuracy of LCA, as well as expand and accelerate discussions aimed at building a sustainable society from a water use perspective. An example of such an approach is a study that, with regard to certain water use, defines the time until the production of the relevant product is impeded without it as an "acceptable delay" and determines the flow of water use based on this time (Uchida 2018).

Phosphorus is one of the three major nutrients needed by plants, and it must be continually supplied to the soil as fertilizer for consistent agricultural production. Proven reserves of phosphate rock worldwide are 67 billion tons, while global production in 2012 was 0.21 billion tons (U.S. Department of the Interior 2013), and a simple calculation of the ratio of reserves to production yields more than 300 years. However, considering increases in global population and food production, as well as demand from other industries such as steel manufacturing, a peak in the mining of high-quality phosphate rock could be reached at a quite an early stage. So, to cope with associated prices increases, it will become necessary to consider the recycling of phosphorus resources. Our understanding of the material flow of phosphorus is improving (Matsubae-Yokoyama et al. 2009), and research is expected to progress in areas such as recycling technology and efficient methods of use. The results of assessments of environmental load that consider phosphorus resource consumption as an environmental impact category based on material flow data are expected to serve as important auxiliary information in this research.

6.5 Closing

The results of environmental impact assessments tend to be misunderstood, and so, communication is an important element in the dissemination of results. Care must be taken to ensure that determining environmental loads is not interpreted as asserting that the target of the assessment is harmful to the environment. Also, there is a limit to the reliability and certainty of the data, and the importance of the absolute values also varies depending on the category. Decisions weighing the pros and cons of introducing and directing technological improvements should be made comprehensively by examining the target's merits and demerits, and considering the specific impacts that the results have on society, rather than simply comparing the size of the figures.

Environmental science is a new discipline, and the same is true of LCA. The situation surrounding the global environment is predicted to change constantly over the coming decades and centuries, and remarkable progress in assessment methods that goes beyond the developments mentioned here is possible. One of the dreams of those involved in this field is the completion of a "unified theory" of environment and economy in environmental impact assessment and a "grand unified theory" that includes society—similar to the four fundamental forces of physics.

References

Bradley MD (1975) Future opportunities for open space. Landsc Plann 2:13–22
Federal Office for the Environment FOEN (2009) The ecological scarcity method – eco-factors 2006. FOEN, Bern
Gleeson T, Wada Y, Bierkens MFP, van Beek LPH (2012) Water balance of global aquifers revealed by groundwater footprint. Nature 488:197–200
Ho DP, Ngo HH, Guo W (2014) A mini review on renewable sources for biofuel. Bioresour Technol 169:742–749
Li C, Frolking S, Frolking TA (1992) A model of nitrous oxide evolution from soil driven by rainfall events: 1. Model structure and sensitivity. J Geophys Res 97:9759–9776
Maeda M (2008) Model ni yoru dojou, noukouchi, ryu-iki ni okeru chisso doutai no rikai: 2. hatake hojo ni okeru houkatsuteki chisso doutai model no kaihatsuto riyou no tameno kiso.n. Dojo Hiryogaku Zasshi 79:89–99. in Japanese
Matsubae-Yokoyama K, Kubo H, Nakajima K, Nagasaka T (2009) J Ind Ecol 13:687–705
Morimoto S, Miyamoto K (2009) A current review of social impact assessment on sustainable biomass/biofuel development. J Jpn Inst Energy 88:133–139
Oki T, Kanae S (2007) Current situation and future perspectives on global hydrologic cycles, water balances, and world freshwater resources. J Geogr 116:31–42
Pfister S, Koehler A, Hellweg S (2009) Assessing the environmental impacts of freshwater consumption in LCA. Environ Sci Technol 43:4098–4104
Rehl T, Lansche J, Müller J (2012) Life cycle assessment of energy generation from biogas—attributional vs. consequential approach. Renew Sustain Energy Rev 16:3766–3775
Ridoutt BG, Sanguansri P, Nolan M, Marks NJ (2012) Meat consumption and water scarcity: beware of generalizations. Cleaner Prod 28:127–133
Shirato Y, Hakamata T, Taniyama I (2004) Modified rothamsted carbon model for andosols and its validation: changing humus decomposition rate constant with pyrophosphate-extractable Al. Soil Sci Plant Nutr 50:149–158
Stoeglehner G, Edwards P, Daniels P, Narodoslawsky M (2011) The water supply footprint (WSF): a strategic planning tool for sustainable regional and local water supplies. J Cleaner Prod 19:1677–1686
Therivel R (1993) Systems of strategic environmental assessment. Environ Impact Assess Rev 13:145–168
U.S. Department of the Interior (2013) U.S. geological survey: mineral commodity summaries 2013. US Geol Surv, Reston: 119
Uchida S (2018) Regional analyses of water use in Japanese paddy rice cultivation using modified water footprint indexes. Asia-Pac J Reg Sci. (online first)
Uchida S, Hayashi K (2012) Comparative life cycle assessment of improved and conventional cultivation practices for energy crops in Japan. Biomass Bioenergy 36:302–315
Uchida S, Mizunoya T, Higano Y (2008) A dynamic evaluation of policies to promote the use of the potential energy in wastes. Stud Reg Sci 38:339–350
Uchida S, Hayashi K, Sato M, Hokazono S (2010a) Construction of agri-environmental data using computational methods: the case of life cycle inventories for agricultural production systems. In: Prado HAD, Luiz AJB, Filho HC (eds) Computational methods applied to agricultural research: techniques and advances. IGI Global, Hershey, pp 412–433
Uchida S, Hayashi K, Watanabe T, Sugiura R (2010b) Evaluating the regional biomass utilization model by scenario-based comparative life cycle assessment. In: Proceedings of the workshop of research results in rural biomass research project supported by Ministry of Agriculture, Forestry and Fisheries of Japan: III–10
Vanham D, Bidoglio G (2013) A review on the indicator water footprint for the EU28. Ecol Indic 26:61–75
Whitaker J, Ludley KE, Rowe R, Taylor G, Howard DC (2010) Sources of variability in greenhouse gas and energy balances for biofuel production: a systematic review. Glob Change Biol Bioenergy 2:99–112

Chapter 7
A Dual Input-Output Approach for Optimal Tax-Subsidy Policy to Reduce Greenhouse Gas and Air Pollutants Emission: Comparison in 1997 and 2000

Takeshi Mizunoya, Noriko Nozaki, Rajeev Kumar Singh, Ailyn Rojas Cabrera, and Helmut Yabar

Abstract In this study, we present an expanded input-output model, which determines the optimal level of economic activities and their lowest emission of air pollutants. The objective is to maximize the welfare function with respect to the emission taxes, being subject to the emission standards of the air pollutants. By making the emission standards strict step by step, we can analyze the feasibility and implementation potential of reduction scenarios by the numerical simulations. The model is applied to the Japanese economy by comparing the simulation results for the years 1997 and 2000. This comparison reveals the types of changes experienced by the Japanese economy in order to achieve the GHG emission reduction target of the Kyoto Protocol. It also reveals the alternative pollutant reduction policy in these years. We analyze one type of greenhouse gas and two types of air pollutants which are CO_2, SOx, and NOx generated by the consumption of fossil fuels and ignore other anthropogenic greenhouse gases. The prices change so as to reflect social costs of the air pollutants through the optimal emission taxes. We consider that these optimized emission taxes are collected by the government and used for the running cost of abatement industries. The taxes will also help to subsidize industries whose activities become unsustainable due to the introduction of the emission tax, in order to avoid generation of idle capital and unemployment labor.

Keywords Greenhouse gas · Air pollutants · Environmental policy · Computer simulation · Extended I-O analysis · Quasi-market

---

T. Mizunoya (✉) · N. Nozaki · A. R. Cabrera · H. Yabar (✉)
University of Tsukuba, Tsukuba-shi, Ibaraki, Japan
e-mail: mizunoya.takeshi.ff@u.tsukuba.ac.jp; yabar.mostacero.h.ke@u.tsukuba.ac.jp

R. K. Singh
Institute for Global Environmental Strategies, Hayama, Kanagawa, Japan

K. Nakayama, Y. Miyata (eds.), *Theoretical and Empirical Analysis in Environmental Economics*, New Frontiers in Regional Science: Asian Perspectives 34, https://doi.org/10.1007/978-981-13-2363-8_7

7.1 Introduction

A major characteristic of the air pollution problem in recent years is that it is caused not by specific pollution sources but by substances emitted from the whole socioeconomic activities. Therefore, it is difficult to control the discharge by the conventional direct regulation policy. In addition to that, the causative substances are not only those that directly affect human health, such as SOx and NOx, but also a variety of substances that indirectly influence human health and the environment through climate change like CO_2. Toward this end it is necessary to have simultaneous control of multiple pollutants. Particularly, climate change affects the survival of human race, and the Parties to the Framework Convention on Climate Change signed the "Kyoto Protocol" in December 1997 to address this challenge. Japan, as a party to the Kyoto Protocol, shall ensure that the aggregate anthropogenic CO_2 equivalent emissions of greenhouse gases such as CO_2, CH_4, etc. do not exceed 94% of the 1990 level in the commitment period 2008–2012.

In this study, we present an expanded input-output model which determines the optimal level of economic activities and their optimal emission of air pollutants so as to maximize the welfare function with respect to the emission taxes, being subject to the emission standards of the air pollutants. Making the emission standards strict step by step, we analyze the feasibility and implementability of reduction scenarios by the numerical simulations. The model is applied to the Japanese economy as of 1997 and 2000. And we compare simulation results for both years. This comparison reveals what kind of change has occurred in the Japanese economy in order to achieve the reduction target of the Kyoto Protocol and what kind of difference should be made in pollutant reduction policy in these years. We analyze one type of greenhouse gas and two types of air pollutants which are CO_2, SOx, and NOx generated by the consumption of fossil fuels and ignore other anthropogenic greenhouse gases.

Our basic point of view in this paper is that the emission taxes alone are not so effective for sufficient reduction in the emission of greenhouse gases and air pollutants contrary to the advocacy of it. The taxes are only effective and useful in that they adjust economic activities within domestic markets so as to reflect the social costs of the activities or else they adjust the activities in international markets either by giving advantage to developing countries through exemption or remittal of the taxes and/or by making negative transfers to the developed countries provided that the taxes are built into the total environmental economic system being together with, e.g., the quasi-markets of emission permits (see Higano 1988, 1997).

7.2 Construction of Simulation Model

We formulate a model of the total environmental economic system that controls the air pollutants emitted by not only industries but also households. The fundamental nature of the model is the *nonlinear* dual system of the input-output analysis. The taxes are levied on the emission of air pollutants caused by both production and consumption activities based upon the "Producers pay" principle. The prices change so as to reflect social costs of the air pollutants through the optimal emission taxes. We consider that these optimized emission taxes are collected by the government and used for the running cost of abatement industries and for subsidizing industries whose activities become unsustainable due to the introduction of the emission tax, in order to avoid generating of idle capital and unemployment labors.

We set Simulation Cases as follows:

[Basic Case]: Basic Case refers to data of the Japanese economy in 1997 and 2000. Only the level of welfare is calculated. There is no optimization and there are no emission taxes.

[Case 0]: We maximize the welfare with respect to the emission taxes, being subject to the emission standards which are the actual levels of emission in the Basic Case.

[Case *n*]: We further optimize the welfare, making the emission standards strict by n % step by step.

7.2.1 Commodity Flow Balance in the Usual Industries

Each usual industry would produce enough to meet all of demand. In our model, e and m are exogenous variables.

$$A_{11}X_1 + A_{12}X_2 + c + g + I + e - m - X_1 \leq 0 \tag{7.1}$$

A_{11}: an input-output coefficient matrix between the usual industries
A_{12}: an input coefficient matrix from the usual industries to the air pollutant abatement industries
X_1: a column vector of the total production of the usual industries
X_2: a column vector of the activity level of the abatement industries
c : a column vector of the consumption
g: a column vector of the government expenditure
I : a column vector of gross investment demand
e: a column vector of export
m: a column vector of import

7.2.2 Commodity Flow Balance in the Abatement Industries

The total emissions consist of emissions from usual industries, abatement industries, and final demand. Further, the net emissions are calculated by subtracting the amount of pollutants abated by abatement industry from total emissions. In our model, A_{21}, A_{22}, and A_{23} are exogenous variables.

$$A_{21}X_1 + A_{22}X_2 + A_{23}\left(l_1c + l_1g\right) - X_2 \equiv Z \tag{7.2}$$

A_{21}: an emission coefficient matrix of the usual industries
A_{22}: an emission coefficient matrix of the abatement industries
A_{23}: an emission coefficient matrix of consumption and government expenditure
l_1: an aggregation row vector whose elements are all one
Z: a column vector of net emission of air pollutants

7.2.3 Value Flow Balance in the Usual Industries

Assuming that the producers shall internalize all the social costs of the emission of air pollutants, we specify the following value flow balance equations of the usual industries. In addition, in our model, v_1 is an exogenous variable.

$$pA_{11}\tilde{X}_1 + \tau A_{21}\tilde{X}_1 + v_1\tilde{X}_1 + \tau A_{23}\left(\widehat{c} + \widehat{g}\right) = p\tilde{X}_1 + \tau_s^u \tag{7.3}$$

p: a row vector of price
τ: a row vector of emission tax
v_1: a row vector of the rate of value added of the usual industries
τ_s^u: a row vector of the subsidy for usual industries to sustain production activity
$\tilde{X}_1$: a diagonal matrix whose diagonal elements are the elements of X_1
$\widehat{c}$: a diagonal matrix whose diagonal elements are the elements of c
$\widehat{g}$: a diagonal matrix whose diagonal elements are the elements of g

7.2.4 Value Flow Balance in the Abatement Industries

Also, assuming that the abatement activities shall be priced by the optimized emission taxes and the government makes the transfers of the rated values to the abatement industries, we specify the following value flow balance equations of the abatement industries. In our model, v_2 is an exogenous variable.

$$pA_{12}\tilde{X}_2 + \tau A_{22}\tilde{X}_2 + v_2\tilde{X}_2 = \tau\tilde{X}_2 + \tau_s^a \tag{7.4}$$

v_2: a row vector of the rate of value added of the abatement industries
τ_s^a: a row vector of the subsidy for abatement industries to sustain abatement activity

7.2.5 Disposal Income of Household

The disposal income is the income that deducted the direct tax from the personal income. τ_0, τ_I, and δ are exogenous variables in our model.

$$Y_d = (1 - \tau_0)(1 - \tau_I - \delta)(v_1 X_1 + v_2 X_2) \quad (7.5)$$

Y_d: the disposable income of the household sector
τ_0: the rate of the direct tax to personal income
τ_I: the average rate of the indirect tax to the GDP
δ: the average depreciation rate to the GDP

7.2.6 Tax Revenue

The total tax revenue of the government consists of direct tax revenue, indirect tax revenue, and the total emission tax revenue.

$$T = \tau_0 (1 - \tau_I - \delta)(v_1 X_1 + v_2 X_2) + \tau_I (v_1 X_1 + v_2 X_2) + \tau_r \quad (7.6)$$

$$\tau_r = \tau A_{21} X_1 + \tau A_{22} X_2 + \tau A_{23} (c + g) \quad (7.7)$$

T: the total tax revenue of the government
τ_r: the total emission tax revenue

7.2.7 Investment and Saving

The net investment is equal to all of saving. In our model, D, r, and s_g are exogenous variables.

$$I_n + l_1 (e - m) - S + S_g \quad (7.8)$$

$$I = r (I_n + D) \quad (7.9)$$

$$D = \delta (v_1 X_1 + v_2 X_2) \quad (7.10)$$

$$S_g = s_g T \quad (7.11)$$

S: a saving of the household sector
S_g: a government saving
I_n: the total amount of the net investment
D: the total depreciation of the capital

r: a given row vector of the rates of investment demand for each industry to the total investment
s_g: the propensity to saving of the government

7.2.8 Government Expenditure

The government pay out the total tax revenue to government consumption, government saving, and the countermeasure-related cost for reduction of air pollutants. c_g is an exogenous variable in our model.

$$T = g_t + S_g + \tau_e \tag{7.12}$$

$$g = c_g g_t \tag{7.13}$$

And in this model, the countermeasure-related cost for reduction of air pollutants consists of the running cost of the abatement industries and the subsidies for industries to sustain their activities.

$$\tau_e = \tau X_2 + \tau_s^u l_2 + \tau_s^a l_3 \tag{7.14}$$

c_g: a column vector of the propensity to consume of the government
g_t: the total government consumption
τ_e: the countermeasure-related cost for reduction of air pollutants (in this paper, we express this cost as *CMR* cost in following sentence)
l_2: aggregation row vectors whose elements are all one
l_3: aggregation row vectors whose elements are all one

7.2.9 Consumption Level and Saving Level of Household

The household pays out disposable income for consumption and saving at fixed rates of α and β, respectively. In our model, α and β are exogenous variables.

$$p\tilde{c} = \alpha Y_d \tag{7.15}$$

$$S = \beta Y_d \tag{7.16}$$

$$\alpha l_2 + \beta = 1 \tag{7.17}$$

α: a column vector of the propensity to consume of the household in which$\alpha \equiv (\alpha_1, \ldots, \alpha_{32})$
β: the propensity to saving of the household

7.2.10 The Total Emission Standards

The total emission standards of Case n are given as follows. In our model, $\overline{Z}$ and n are exogenous.

$$Z \leq \overline{Z}^{*} (100 - n) / 100 \tag{7.18}$$

$\overline{Z}$: the actual level of emission of air pollutants in 1997
n: the reduction rate in the emission in %

7.2.11 Objective Function

Assuming the Cobb-Douglas utility function in the variables of consumption and saving, the nonlinear optimization problem of Case ($n = 0$, 5, 10, and 15) are given as follows.

[Optimization of Case n]

$$\max_{\{X_1, X_2, p, \tau, c, S, S_g\}} c_1^{\alpha_1} c_2^{\alpha_2} \ldots c_{32}^{\alpha_{32}} S^{\beta} \tag{7.19}$$

subject to Eqs. (7.1), (7.2), (7.3), (7.4), (7.5), (7.6), (7.7), (7.8), (7.9), (7.10), (7.11), (7.12), (7.13), ((7.14), (7.15), (7.16), (7.17), and (7.18)

In this paper, we do not consider any further which system implements the reduction scenario of air pollutant emissions, nor how to audit the system. However, here we would like to point out a possibility and the robustness of the simulation. Namely, one of the means is given by quasi-markets of emission permits that are issued (e.g., by the government) based on this type of simulation. The optimized emission taxes, τ, reflect the social costs of the air pollutants in terms of abatement opportunity cost. The maximization simulates the general equilibrium of the markets (including artificial quasi-markets), in which distortions due to the emission in the production and consumption are corrected by the allocation of emission permits through the equilibrium of the quasi-markets being linked to existing markets. Note here also that the equilibrium is a *general* equilibrium, in that the revenue from the sale of the emission permits, being combined with revenues from the direct and indirect taxes, is spent by the government for its consumption, saving, or subsidy for industrial activities, and the taxes, τ, are shadow prices of the emission permits which should have been realized in the quasi-markets. This implies that if the government would issue the emission permits in an amount equal to the reduction specified by the emission standard, then the optimized rates of tax, τ, would be realized in the quasi-market as prices.

7.3 The Specification of Simulation Model

In the simulation, we analyze the three types of air pollutants listed in Table 7.1, in which carbon dioxide is the main greenhouse gas. And the cases of the simulation in each year are summarized in Tables 7.2 and 7.3.

Tables 7.4 and 7.5 show industry names in production sector and the emission amount of greenhouse gas and air pollutants from each sector in 1997 and 2000. It is obvious that there are two big industries overwhelmingly responsible for greenhouse gas and air pollutant emissions in both years, i.e., industries #18 (electricity, gas) and #23 (transport), which are arterial industries. The next largest emitters in are industries #8 (non-metallic mineral products) and #9 (iron and steel) and final demand in bot year. We can see the CO_2 emission from the #9 (iron and steel) industry increased by 16.6% from 1997 to 2000. And this industry became the third largest emission industry from the fifth largest emission industry of CO_2 in the same time span. We should note industry #1 (agriculture) emits a large amount of NOx and that amount is third largest in both years. The emission standard of Case 0 is based upon the estimation of the total pollutants emitted in 1997 and 2000, respectively (Tables 7.2 and 7.3).

We determined that the abatement activities in each year have the same internal input structure as that of industry #12, general machinery, in the same year because we considered that, assuming the abatement industries abate the air pollutant-utilizing machines, the industrial structure would be similar to that of industry #12. And further assuming that the abatement costs of CO_2, SOx, and NOx are 84 Japanese yen/kg (Society for the System Research of the Greenhouse Gas Effects on the Global Environment of The Environmental Agency, Government of Japan,

Table 7.1 Classification of pollutants

Index	Air pollutants
1	Carbon Dioxide (CO_2)
2	Sulfur Oxides (SOx)
3	Nitrogen Oxides (NOx)

Table 7.2 Case of the simulation (for 1997)

	Reduction rate (%)	(Basic)	0	5	10	15
Upper constraint On the emission of	CO_2 (1000 million t)	1.308	1.308	1.243	1.177	1.112
	SOx (million t)	1.906	1.906	1.810	1.715	1.620
	NOx (million t)	3.881	3.881	3.687	3.493	3.299

Table 7.3 Case of the simulation (for 2000)

	Reduction rate (%)	(Basic)	0	5	10	15
Upper constraint On the emission of	CO_2 (1000 million t)	1.289	1.289	1.225	1.160	1.096
	SOx (million t)	1.799	1.799	1.709	1.619	1.529
	NOx (million t)	3.600	3.600	3.420	3.240	3.060

Table 7.4 Classification of industry and emission of greenhouse gas and air pollutants in Japan (1997)

Industry	Pollutants		
	CO_2 (1000 t)	SOx (t)	NOx (t)
1 Agriculture	17,509	55,880	297,804
2 Mining	916	590	1468
3 Food products	15,507	79,018	16,943
4 Textiles and clothing	3827	16,582	5003
5 Paper and wooden products	30,833	87,478	32,103
6 Chemical products	49,806	107,161	76,613
7 Coal and petroleum products	81,032	78,285	74,135
8 Nonmetallic mineral products	94,777	43,384	151,988
9 Iron and steel	88,379	86,351	84,771
10 Nonferrous metals	7244	18,338	19,518
11 Fabricated metal products	3513	2314	3297
12 General machinery	5053	6595	5076
13 Electric machinery	4409	5163	4371
14 Transport equipment, automobiles	5713	9809	7451
15 Precision apparatus	362	554	345
16 Other manufactures	5674	20,707	7689
17 Building and construction	15,626	8907	12,447
18 Electricity, gas	419,026	333,657	312,770
19 Water	32,109	32,351	37,006
20 Wholesale and retail service	16,746	32,134	12,428
21 Finance and insurance	900	341	639
22 Real estate services	4907	401	4855
23 Transport	183,650	690,351	2,281,925
24 Communications	1462	1693	798
25 "Public works administrative organizations" or "public works"	14,983	20,021	27,974
26 Education and research services	8077	27,606	11,701
27 Medical, health, and social insurance services	11,741	40,130	17,010
28 Other public services	1891	4116	1879
29 Miscellaneous business services	7522	16,371	7472
30 Miscellaneous personal services	23,102	50,281	22,950
31 Office supplies	2473	4847	10,344
32 Not elsewhere classified	8364	16,396	34,991
Total emitted by the production sector	1,167,133	1,897,811	3,585,765
Total emitted by final demand	140,934	7943	295,138
Total emitted	1,308,067	1,905,754	3,880,904

The second interim session 1994), 332 yen/kg, and 2405 yen/kg (Society for the Research of the Global Environmental Economics, Experience of Public Bads in Japan: Diseconomies of the Economy which does not Consider the Environment 1991), respectively, we have estimated the input coefficients of the abatement

Table 7.5 Classification of industry and emission of greenhouse gas and air pollutants in Japan (2000)

Industry	Pollutants		
	CO_2 (1000 t)	SOx (t)	NOx (t)
1 Agriculture	16,884	111,823	201,958
2 Mining	740	1207	5011
3 Food products	14,848	49,516	19,672
4 Textiles and clothing	3551	10,917	7872
5 Paper and wooden products	19,226	42,578	32,571
6 Chemical products	50,536	73,417	70,740
7 Coal and petroleum products	40,725	58,533	66,925
8 Nonmetallic mineral products	68,159	28,539	149,380
9 Iron and steel	164,375	55,910	76,396
10 Nonferrous metals	5488	14,168	12,825
11 Fabricated metal products	5098	2782	6729
12 General machinery	3860	3550	7821
13 Electric machinery	5892	4910	8015
14 Transport equipment, automobiles	6886	5709	9843
15 Precision apparatus	533	401	579
16 Other manufactures	8024	15,952	13,380
17 Building and construction	14,310	11,210	121,489
18 Electricity, gas	378,565	226,467	264,485
19 Water	31,803	24,110	24,483
20 Wholesale and retail service	12,941	30,925	12,128
21 Finance and insurance	1088	291	1413
22 Real estate services	3238	3364	4724
23 Transport	211,265	930,629	2,029,836
24 Communications	1620	2037	5469
25 "Public works administrative organizations" or "public works"	10,769	22,340	25,729
26 Education and research services	11,223	14,804	21,401
27 Medical, health, and social insurance services	13,141	20,160	22,975
28 Other public services	1195	1095	1612
29 Miscellaneous business services	6261	6279	13,529
30 Miscellaneous personal services	20,994	13,537	30,815
31 Office supplies	0	0	0
32 Not elsewhere classified	1728	3204	6968
Total emitted by the production sector	1,134,967	1,790,364	3,276,771
Total emitted by final demand	154,347	8682	323,357
Total emitted	1,289,314	1,799,046	3,600,128

activities. The input coefficients of the usual industries, A_{11}, and the abatement industries, A_{12}, in each year are given in Tables 7.16, 7.17, 7.18, 7.19, 7.20, and 7.21 in the Appendix (Economic Planning Agency of Japan 1999), respectively. Also, assuming that the air pollutants emitted by the abatement activities are proportional to those emitted by industry #12 (general machinery), we have estimated the emission coefficients of the abatement activities. The emission coefficients, A_{21}, A_{22}, and A_{23}, are given in Tables 7.21, 7.22, 7.23, and 7.24 (Yoshioka et al. 1996) in the Appendix. The commodity of the 32nd industry is taken as a numeraire due to the Walras's Law (i.e., $p_{32} \equiv 1$). Other exogenous parameters are listed in the Appendix (Tables 7.25, 7.26, 7.27, and 7.28).

In the simulation, taking account of the production factors such as labor and capital which are unable to change drastically in the static sense, we especially add the constraint that the range of changes in production must be in the range of $-20\% \sim +10\%$ of the total production of the usual industries in the Basic Case. And in order that the emission tax may make preferably the earmarked tax to reduce the air pollutant emission, we also add the constraint that the ratio of the portion of the usual tax revenue transferred to the *CMR* cost/total usual tax revenue must be less than 5%.

We ran this simulation with "LINGO" which is the computer software for operations research released by LINDO SYSTEMS.

7.4 Simulation Result

7.4.1 The Optimized Taxes and Subsidies

The optimized taxes for each year are summarized in Tables 7.6 and 7.7. Here, note that the air pollutants are produced jointly by the production activities with the fixed coefficients of the Harrod-Domar-Leontief type. So, the optimized tax(es) which shall be charged on each emission of air pollutants can be put together by each industry into an *ad valorem* tax that will be charged on the shipped value of the industry. They are summarized in Tables 7.8 and 7.9, respectively. The optimized ad valorem tax rates in 1997 are less than 2.0% in many industries in all cases, while in nine industries, e.g., #1 (agriculture), #5 (paper and wooden products), #6 (chemical products), #7 (coal and petroleum products), #8 (nonmetallic mineral products), #9 (iron and steel), #18 (electricity, gas), #19 (water), and #23 (transport), there are several cases in which they exceed 2%. Industries #18 (electricity, gas) and #23 (transport) show especially high rates. On the other hand, the optimized ad valorem tax rates in 2000 were a little higher than in 1997 in average. Industries #1 (agriculture), #8 (nonmetallic mineral products), #9 (iron and steel), #18 (electricity, gas), and #23(transport) showed more than 10% increase in some cases. Especially, #18 (electricity, gas) in Case 0 and #23 (transport) in Cases 10 and 15 are imposed higher than 30% of tax. This result shows that the social cost of greenhouse gas and air pollutants which should be paid by these industries in 2000 was very large and its value became much larger than the value in 1997 in just 3 years (e.g., the social cost that # 23 (transport) should pay in Case 15 has increased 4.6 times over 3 years).

Table 7.6 Optimized taxes in 1997

Reduction rate (%)	0	5	10	15
CO_2 (thousand yen/t)	3	0	0	0
SOx (thousand yen/t)	0	0	0	5762
NOx (thousand yen/t)	233	1653	2425	0

Table 7.7 Optimized taxes in 2000

Reduction rate (%)	0	5	10	15
CO_2 (thousand yen/t)	17	0	0	0
SOx (thousand yen/t)	0	0	0	8316
NOx (thousand yen/t)	842	5830	7266	8265

Table 7.8 Total emission tax per shipping million in 1997 (10,000 yen)

Reduction rate (%)		0	5	10	15
Industry	1	0.88	3.39	4.98	2.22
	2	0.17	0.12	0.18	0.17
	3	0.13	0.07	0.10	1.11
	4	0.14	0.09	0.12	0.98
	5	0.62	0.30	0.44	2.86
	6	0.67	0.46	0.68	2.26
	7	2.26	0.98	1.43	3.59
	8	3.59	2.60	3.81	2.59
	9	1.35	0.61	0.90	2.17
	10	0.41	0.46	0.68	1.52
	11	0.08	0.03	0.05	0.08
	12	0.04	0.02	0.03	0.09
	13	0.03	0.01	0.02	0.05
	14	0.04	0.03	0.04	0.12
	15	0.03	0.01	0.02	0.07
	16	0.06	0.04	0.06	0.37
	17	0.06	0.02	0.03	0.05
	18	7.77	2.77	4.06	10.29
	19	2.16	1.16	1.70	3.53
	20	0.06	0.02	0.03	0.20
	21	0.01	0.00	0.00	0.01
	22	0.02	0.01	0.02	0.00
	23	3.15	10.47	15.35	11.04
	24	0.03	0.01	0.01	0.06
	25	0.14	0.12	0.17	0.29
	26	0.13	0.09	0.13	0.72
	27	0.13	0.09	0.13	0.72
	28	0.10	0.05	0.07	0.35
	29	0.04	0.02	0.03	0.16
	30	0.13	0.06	0.09	0.47
	31	0.58	0.94	1.38	1.53
	32	0.58	0.94	1.38	1.53

Table 7.9 Total emission tax per shipping million in 2000 (10,000 yen)

Reduction rate (%)		0	5	10	15
Industry	1	3.17	8.19	10.21	18.09
	2	1.21	2.12	2.64	3.73
	3	0.69	0.29	0.37	1.48
	4	0.94	0.65	0.81	2.20
	5	2.37	1.28	1.59	4.19
	6	3.50	1.58	1.97	4.58
	7	5.74	3.01	3.75	8.01
	8	15.28	10.41	12.97	17.59
	9	16.58	2.59	3.23	6.39
	10	1.69	1.22	1.52	3.65
	11	0.68	0.29	0.36	0.59
	12	0.25	0.16	0.20	0.33
	13	0.20	0.09	0.11	0.20
	14	0.29	0.13	0.17	0.30
	15	0.24	0.09	0.11	0.21
	16	0.45	0.24	0.30	0.75
	17	0.45	0.92	1.14	1.42
	18	34.37	7.99	9.96	21.10
	19	7.24	1.85	2.31	5.22
	20	0.24	0.07	0.09	0.37
	21	0.05	0.02	0.03	0.04
	22	0.09	0.04	0.05	0.10
	23	11.03	24.70	30.79	51.17
	24	0.14	0.14	0.18	0.28
	25	0.56	0.41	0.52	1.10
	26	0.57	0.34	0.43	0.83
	27	0.55	0.30	0.38	0.81
	28	0.51	0.22	0.28	0.53
	29	0.15	0.10	0.13	0.22
	30	0.65	0.31	0.38	0.63
	31	0.00	0.00	0.00	0.00
	32	0.83	0.96	1.20	2.00

It is interesting to observe that the taxes are charged on the shipment of industries #7 (coal and petroleum products), #8 (nonmetallic mineral products), #18 (electricity, gas), #19 (water), and #23(transport) with relatively high rates even in the 0% reduction case. This means the distortion caused by these industries due to social costs of greenhouse gas and air pollutants is larger than ever considered. One reason is that they emit the largest amount of CO_2 (in the cases of industries #18 (electricity, gas) and #23 (transport) – therefore the emission coefficients are relatively large) and they supply essential inputs for other industries. The other reason is that their emission coefficients are relatively large, though the total

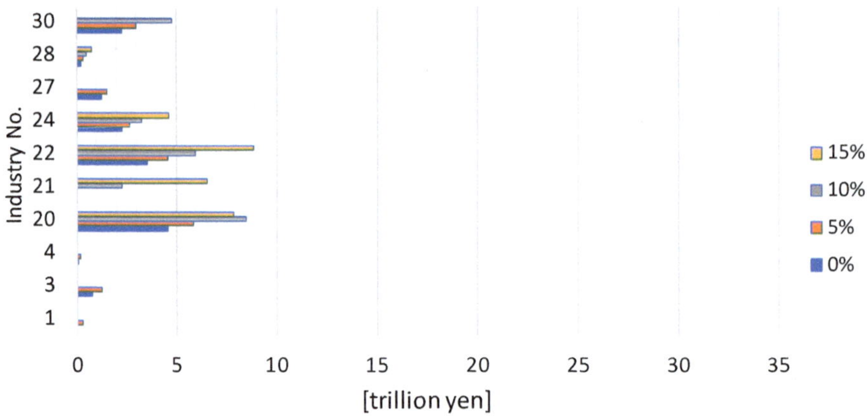

Fig. 7.1 Subsidies for usual industries to continue production activity in 1997

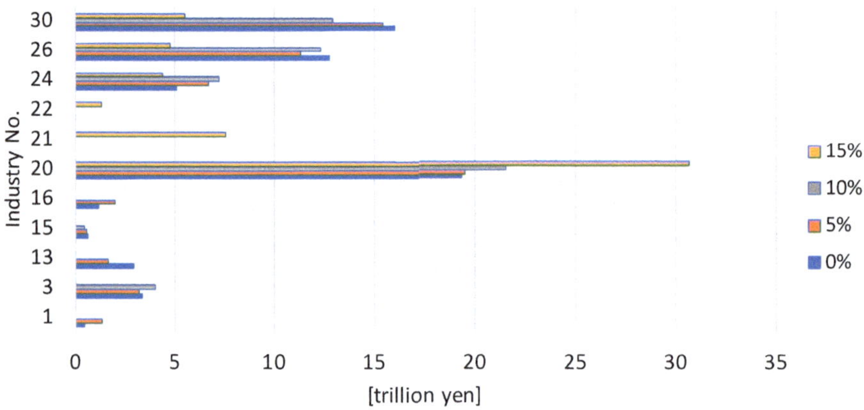

Fig. 7.2 Subsidies for usual industries to continue production activity in 2000

emissions are not so large (in the cases of industries #7 (coal and petroleum products), #8 (nonmetallic mineral products), and #19 (water)).

Subsidies for usual industries to sustain production activity in each year are shown in Figs. 7.1 and 7.2. Optimized subsidies in 1997 and 2000 are found in only 10 and 11 industries, respectively, including #20 (wholesale and retail service) and #22 (real estate services) which produce high added value. However, subsidy amount for each industry in 2000 is so much larger than in 1997 (e.g., subsidy for #20 (wholesale and retail service) in Case 15 in 2000 is 3.9 times larger than that in 1997). This result shows that introduction of emission tax in 2000 so much affect the income of these industries and the government has to cover this damage with their tax revenue. However, this result also shows that as long as the government introduces pollution emissions taxes and at the same time provides subsidy expenditure as shown here, it can reduce environmental

Table 7.10 Emission tax revenue and expenditure in 1997

Reduction rate (%)	0	5	10	15
(1) Total tax rev. ((2) + (3)) (trillion yen)	109.43	112.83	117.88	121.65
(2) Total usual tax rev. (trillion yen)	95.30	94.19	92.32	88.61
(3) Total emission tax rev. (trillion yen)	14.14	18.63	25.56	33.03
(4) The measure cost for reduction of air pollutants (trillion yen)	14.88	19.61	26.90	34.77
(5) The transfer from total usual tax rev. ((4)–(3))	0.74	0.98	1.35	1.74
The rate of total emission tax rev. to total tax rev. (%)((3)*100/(1))	12.92	16.52	21.68	27.15
The rate to the usual tax rev. of the transfer to the measure cost (%) ((5)*100/(2))	0.78	1.04	1.46	1.96
The rate to the measure cost of the transfer from the usual tax rev. (%)((5)*100/(4))	5.00	5.00	5.00	5.00

Table 7.11 Emission tax revenue and expenditure in 2000

Reduction rate (%)	0	5	10	15
(1) Total tax rev. ((2) + (3)) (trillion yen)	186.61	187.66	185.19	175.37
(2) Total usual tax rev. (trillion yen)	127.42	126.90	125.30	119.30
(3) Total emission tax rev. (trillion yen)	59.19	60.77	59.89	56.07
(4) The measure cost for reduction of air pollutants (trillion yen)	62.30	63.97	63.04	59.02
(5) The transfer from total usual tax rev. ((4)–(3))	3.12	3.20	3.15	2.95
The rate of total emission tax rev. to total tax rev. (%) ((3)*100/(1))	31.72	32.38	32.34	31.97
The rate to the usual tax rev. of the transfer to the measure cost (%) ((5)*100/(2))	2.44	2.52	2.52	2.47
The rate to the measure cost of the transfer from the usual tax rev. (%) ((5)*100/(4))	5.00	5.00	5.00	5.00

impact while maintaining economic activity to some extent. On the other hand, the subsidy is not found in other industries such as industries #18 (electricity, gas) and #23 (transport) in both years. This result shows that introduction of emission tax decreases the income of industries which have comparatively small emission coefficients and high consumption propensities (e.g., industries #20 (wholesale and retail service), #22 (real estate services), and #30 (miscellaneous personal services)) and the industries which have, similarly, comparatively small emission coefficients and supply essential inputs for other industries (e.g., industry #24 (communications)), rather than industries whose emission tax rates are high (e.g., industries #18 (electricity, gas) and #23 (transport)).

The total emission tax revenues in 1997 and 2000 were about 14–33 trillion yen and 56–60 trillion yen, respectively (see Tables 7.10 and 7.11). This result reflects the total social cost of CO_2, SOx, and NOx. From these results we can infer that there is a difference of about 27–42 trillion yen in social cost in both years. It is interesting that the value in 2000 is nearly stable, while the value in 1997 increases as the reduction rate increases. The total emission tax revenue amount to the total tax revenue in 1997 was about 13–27% and that in 2000 was about 31–32%. And those values in 2000 were higher than those in 1997 in each case. The government expenditure for the *CMR* cost exceeds the total emission tax revenue in all cases in both years. In this study, we ran a simulation in which there was no transferred amount from usual tax revenue to the *CMR* cost (in other words, the case that the emission tax is introduced as entire earmarked tax), but we could not obtain feasible solutions in all simulation cases. Furthermore, we also ran a contrary simulation in which the emission tax was introduced not as the earmarked tax to reduce air pollutant emissions but as an ordinary tax. In all of the solutions of this case in both years, the rate of the *CMR* cost to the total tax revenue exceeded 20%, and these solutions were unrealistic. So, in this simulation we assumed that the ratio of the usual tax revenue transferred to the CMR cost to the total usual tax revenue must be less than 5%. In the case of 1997, the stricter the reduction rate became, the larger the transferred amount from usual tax revenue to the *CMR* cost become. However, the ratio of the portion of the usual tax revenue transferred to the *CMR* cost/the *CMR* cost in 1997 was only 1.96% and at maximum level, and it hardly affects usual government expenditures. On the other hand, although the transferred amount from usual tax revenue to the *CMR* cost in 2000 became larger as the reduction rate became larger by 10%, that amount in Case 15 (15% reduction case) was smaller than that in Case 10 (10% reduction case). However, the ratio of the portion of the usual tax revenue transferred to the *CMR* cost/the *CMR* cost in 2000 was 2.52% and at maximum level, and it also hardly affects usual government expenditures. These results show that even if the emission tax is introduced to Japanese economy, the total emission tax revenue cannot cover all of the *CMR* cost in any reduction cases. However, the *CMR* cost can consist with usual tax expenditure as long as we introduce the emission tax as the earmarked tax and allow a small transfer from usual tax revenue to the *CMR* cost (see Tables 7.10 and 7.11).

7.4.2 Gross Emission and Abatement of Greenhouse Gas and Air Pollutants

The total air pollutants emitted and abated in each year are shown in Tables 7.12 and 7.13. In this simulation, the abatement activities are inactive in some cases in both years. Among three kinds of abatement industries, the abatement activity of CO_2, which has smallest abatement cost, was made in Case 10 and Case 15 in 1997 and in Case 10, Case 15, and Case 20 in 2000. The maximum abatement level remained at 4.67% in Case 15 in 2000. This result indicates that in order to achieve this reduction rate, it is necessary to cover the 4.67% reduction of pollutants by emissions trading or to develop abatement technology that can reduce at least this amount of pollutants. Regarding SOx abatement activity, it was attained only in Case 15 in 1997, but the rate was very low (0.74%).

Table 7.12 Emission and abatement in 1997

Reduction rate (%)	(Basic)	0	5	10	15
CO_2 (1000 million t)					
Total emission	1.3081	1.3081	1.2427	1.1960	1.1492
Total abatement	0	0	0	0.0188	0.0374
Net emission	1.3081	1.3081	1.2427	1.1773	1.1119
Abatement rate (%)	0	0	0	1.57	3.25
SOx (million t)					
Total emission	1.9058	1.9058	1.8105	1.7152	1.6320
Total abatement	0	0	0	0	0.0121
Net emission	1.9058	1.9058	1.8105	1.7152	1.6199
Abatement rate (%)	0	0	0	0	0.74
NOx (million t)					
Total emission	3.8778	3.8778	3.6869	3.4928	3.2481
Total abatement	0	0	0	0	0
Net emission	3.8778	3.8778	3.6869	3.4928	3.2481
Abatement rate (%)	0	0	0	0	0

Table 7.13 Emission and abatement in 2000

Reduction rate (%)	(Basic)	0	5	10	15
CO_2 (1000 million t)					
Total emission	1.2893	1.2893	1.2502	1.2032	1.1496
Total abatement		0	0.0253	0.0428	0.0537
Net emission	1.2893	1.2893	1.2248	1.1604	1.0959
Abatement rate (%)		0	2.03	3.56	4.67
SOx (million t)					
Total emission	1.7990	1.7990	1.7091	1.6191	1.5291
Total abatement		0	0	0	0
Net emission	1.7990	1.7990	1.7091	1.6191	1.5291
Abatement rate (%)		0	0	0	0
NOx (million t)					
Total emission	3.6001	3.6001	3.6001	3.2401	3.0601
Total abatement		0	0	0	0
Net emission	3.6001	3.6001	3.6001	3.2401	3.0601
Abatement rate (%)		0	0	0	0

Furthermore, abatement activity of NOx, which has the highest abatement cost, was not attained at all in both years. This result shows that the higher the abatement cost, the more difficult to reduce emissions by methods other than reducing the production of goods in the industrial sector and reducing consumption in the final demand sector. Therefore, in order to reduce emissions of air pollutants while maintaining a certain economic level, it is indispensable to develop low-cost abatement technologies.

7.4.3 Changes in Consumption, Production, and Price

Figures 7.3 and 7.4 show the changing in production of each industry in each year, respectively. The rate of the production change in percentage is in the range of −20% to +10%, which we set as upper and lower limit. In both years, production amount of industries #18 (electricity, gas) and #23 (transport) is drastically reduced to −20% in some cases. This is because the large amount of emission tax was imposed on these industries, and the demand for these industries was decreased due to the introduction of such taxes. The consumption demand changes drastically between industries and the cases in both years (see Figs. 7.5 and 7.6). Especially regarding the increase in consumption demand, it is remarkable in 2000. For example, consumption demand for industry #24 (communications) in Case 10 in 2000 increases by 65%, while consumption demand to industry #18 (electricity, gas) is decreased by 70%. This is due to changes in prices affected by the emission taxes (see Figs. 7.7 and 7.8). It should be noted that by charging the emission taxes which reflect the social costs of the air pollutants, the consumption demand for industries #18 (electricity, gas) and #23 (transport) is drastically decreased in both years, even though the reduction rate is zero (see Figs. 7.5 and 7.6). On the other hand, the consumption demand and the amount of production of the industries with

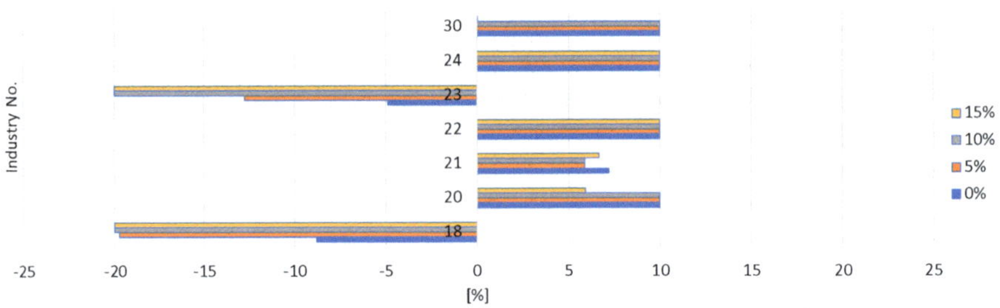

Fig. 7.3 Production in percentage to the basic case in 1997

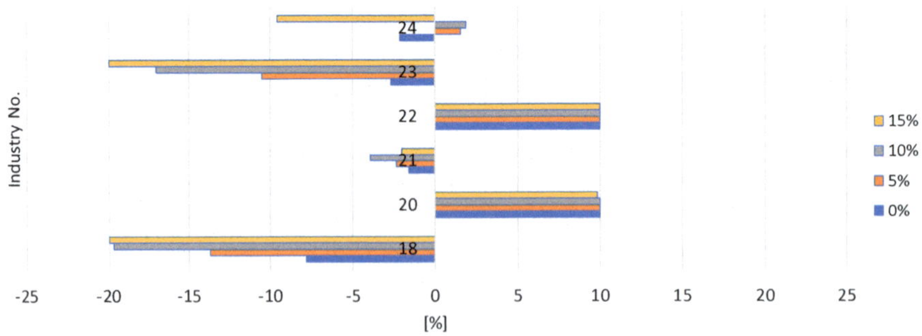

Fig. 7.4 Production in percentage to the basic case in 2000

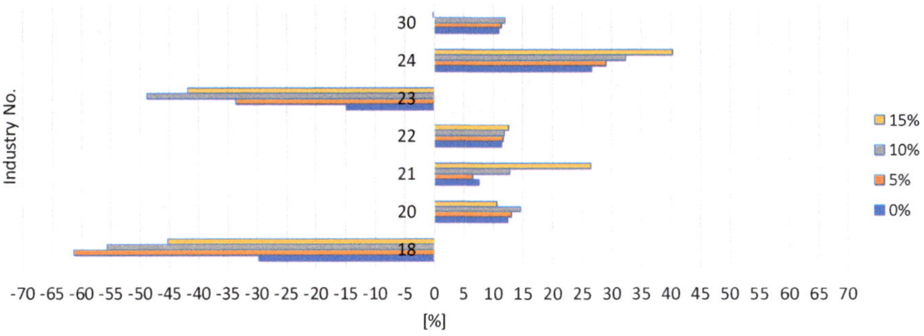

Fig. 7.5 Consumption demand for each industry in percentage to the basic case in 1997

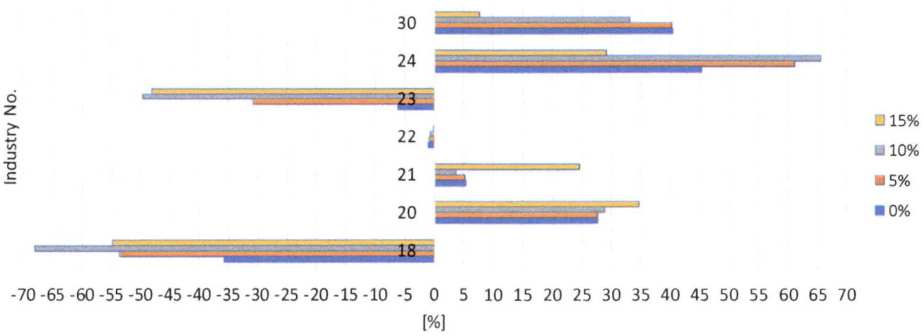

Fig. 7.6 Consumption demand for each industry in percentage to the basic case in 2000

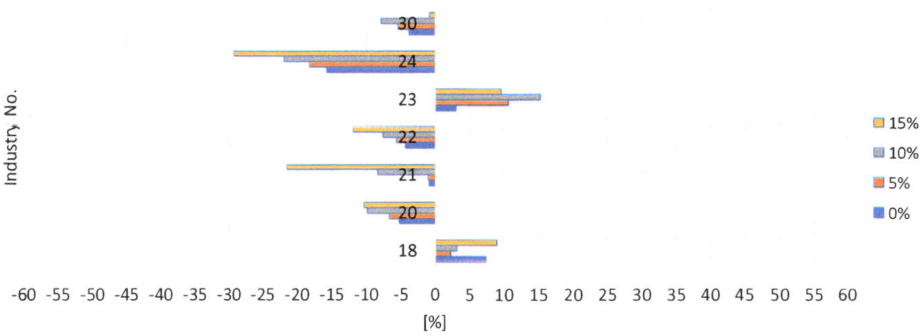

Fig. 7.7 Change in price to the basic case in 1997

large consumption propensities and comparatively small emission coefficients such as industries #20 (wholesale and retail service), #21 (finance and insurance), and #30 (miscellaneous personal services) or industry #24 (communications), which are indispensable to other industries, and which have comparatively small emission coefficients, increased as compared to the basic case, in all cases.

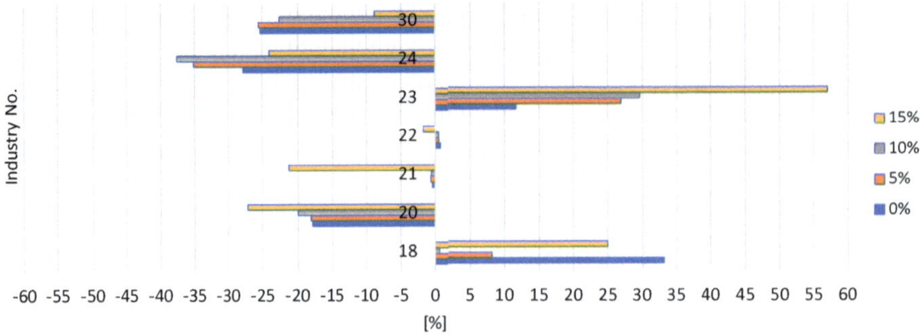

Fig. 7.8 Change in price to the basic case in 2000

Since especially high taxes are imposed on two industries (#18 (electricity, gas), #23 (transport)), which have higher amounts of emissions, the price of commodities in these industries increases as the reduction rate becomes larger. On the other hand, the commodity prices of #20 (wholesale and retail service), #24 (communications), and #30 (miscellaneous personal services) are decreased remarkably by the introduction of the emission tax, as if to neutralize the negative influence of the rise in the two *big* industries, which are #18 (electricity, gas) and #23 (transport). Because of this change in commodity price, the consumption demand and the amount of production of these industries increase, and, on the other hand, the costs of intermediate demand from these industries to others decrease. As the result of this process, the national welfare is maximized (in other words, the loss of national welfare by introduction of the emission tax is minimized) under the strict emission constraints.

It is interesting that the commodity price of #22 (real estate services) increased in some cases in 2000, while that price of the industry in 1997 decreased in all cases. As a result, consumption demand for this industry decreased in some cases in 2000, while that demand for the industry in 1997 increased in all case.

7.4.4 Gross Domestic Product and Welfare

The changes in gross domestic product (GDP) and the welfare (the objective value of this simulation) in 1997 and 2000 are shown in Tables 7.14 and 7.15, respectively. The GDP in 1997 was 507.7 trillion Japanese yen and that in 2000 was 519.5 trillion Japanese yen. By charging the emission taxes to remedy the distortion, the GDP is increased to 541.6 trillion yen in 1997 and to 544.7 trillion yen in 2000. The GDP value in both years decreases gradually as the reduction rate increases by 10% and then drastically decreased in Case 15 to smaller value than that in Basic case.

On the other hand, due to the charge of the emission taxes, the (virtual) optimized value of the welfare increases from 25.0 trillion yen in Basic case to 26.2 trillion yen

Table 7.14 GDP and objective value in 1997

Reduction rate (%)	(Basic)	0	5	10	15
GDP (trillion yen)	507.7	541.6	535.4	524.8	503.7
Objective value (trillion yen)	25.0	26.2	25.7	25.0	23.2

Table 7.15 GDP and objective value in 2000

Reduction rate (%)	(Basic)	0	5	10	15
GDP (trillion yen)	519.5	544.7	542.4	535.6	510.0
Objective value (trillion yen)	24.1	24.8	24.4	23.8	22.8

in Case 0 in 1997. As for 2000, the (virtual) optimized value of the welfare increases from 24.1 trillion yen in Basic case to 24.8 trillion yen in Case 0. In both years, same as GDP value, the welfare value decreases gradually as the reduction rate increases by 10% and then drastically decreased in Case 15 to smaller value than that in Basic case. From these results, it could be considered that even the reduction rate of 15% would be too large to be accepted and a maximum tolerable reduction level could be 10% or so as for the economy of both 1997 and 2000. The tolerable emission reduction level in both years do not reached to the required level to achieve the agreement of the Kyoto Protocol, which is 23.4% less than the actual emission level in 1997.

7.5 Conclusion

In this research we present the model that endogenously determines the optimal taxes on the emission of greenhouse gasses of CO_2, SOx, and NOx due to the consumption of fossil fuel. And we applied the model to the Japanese economy as of 1997 and 2000 to analyze situational changes in these years and propose optimal policy. Through this study, we found that the policies to be implemented to reduce greenhouse gas and air pollutant emissions have changed dramatically in just 3 years from 1997 to 2000. The model is applicable for the situation in which, e.g., the government has to determine what amount of the greenhouse gasses/the air pollutants shall be reduced and how to implement the policy goal. The model is a general equilibrium model in that the government spends the revenue of the taxes for its consumption, investment, and subsidization for industries. So, the model is useful, e.g., when the government has to implement the reduction scenario through "quasi-markets" of the emission permits.

At the end of this research, we would like to emphasize the necessity of rapid and drastic progress in abatement technology. A fairly large amount of R&D investment in abatement technology and related sciences will pay more than ever thought generally.

A.1 Appendix

Table 7.16 Input coefficients of usual industries and other parameters (1997) (no.1)

Industry	1	2	3	4	5	6	7	8	9	10	11	12	13	14	15	16
1	1.176E-01	8.161E-04	2.036E-01	1.148E-02	5.661E-02	1.975E-03	9.729E-05	2.983E-04	9.035E-05	3.327E-04	2.687E-04	2.432E-04	3.194E-04	1.867E-04	2.736E-04	7.483E-03
2	2.067E-07	2.412E-03	3.655E-05	2.595E-05	1.511E-03	4.272E-03	4.251E-01	1.008E-01	1.281E-02	8.843E-02	9.085E-05	2.464E-04	8.505E-05	2.997E-05	2.186E-05	2.609E-03
3	8.289E-02	4.587E-03	1.511E-01	3.608E-03	4.184E-03	9.633E-03	8.398E-04	3.219E-03	1.085E-03	1.420E-03	3.160E-03	1.993E-03	2.554E-03	1.518E-03	2.100E-03	5.460E-03
4	4.631E-03	2.059E-03	9.796E-04	2.481E-01	5.279E-03	1.017E-03	2.767E-04	1.990E-03	4.291E-04	1.331E-03	1.270E-03	8.487E-04	1.445E-03	2.122E-03	1.825E-03	6.310E-03
5	1.124E-02	2.988E-03	2.019E-02	1.054E-02	2.557E-01	1.558E-02	3.262E-04	2.161E-02	4.695E-03	6.685E-03	6.494E-03	3.049E-03	8.322E-03	2.736E-03	1.033E-02	5.578E-02
6	3.961E-02	6.606E-03	9.259E-03	7.356E-02	2.899E-02	2.783E-01	3.283E-03	2.125E-02	5.030E-03	1.959E-02	9.190E-03	6.086E-03	1.210E-02	1.137E-02	8.778E-03	1.010E-01
7	1.757E-02	6.522E-02	4.503E-03	4.594E-03	8.142E-03	4.477E-02	3.611E-02	2.231E-02	2.660E-02	8.554E-03	4.695E-03	2.369E-03	2.848E-03	1.877E-03	2.884E-03	3.936E-03
8	1.480E-03	3.606E-04	6.772E-03	5.519E-04	5.620E-03	5.127E-03	7.743E-04	9.681E-02	4.425E-03	4.001E-03	4.188E-03	4.353E-03	1.014E-02	7.063E-03	1.825E-02	3.460E-03
9	4.231E-05	1.597E-03	2.456E-05	9.380E-05	6.745E-03	1.809E-04	1.230E-03	1.146E-02	4.752E-01	1.113E-02	1.832E-01	6.765E-02	1.573E-02	3.347E-02	1.126E-02	4.444E-03
10	1.047E-05	8.572E-05	1.072E-03	3.686E-05	1.161E-03	3.392E-03	5.927E-05	2.838E-03	1.018E-02	3.112E-01	5.580E-02	2.011E-02	3.595E-02	1.154E-02	2.159E-02	6.655E-03
11	1.772E-03	1.701E-02	2.237E-02	2.784E-03	1.567E-02	1.042E-02	2.375E-03	7.399E-03	6.210E-04	4.207E-03	5.646E-02	3.200E-02	2.083E-02	9.886E-03	1.530E-02	8.230E-03
12	9.531E-03	1.297E-02	5.895E-03	5.176E-03	8.115E-03	9.717E-03	6.855E-03	1.675E-02	7.345E-03	9.685E-03	1.500E-02	2.212E-01	1.901E-02	7.787E-02	1.810E-02	8.509E-03
13	7.871E-04	3.694E-03	8.430E-04	1.041E-03	9.381E-04	1.873E-03	5.856E-04	7.236E-04	2.817E-03	1.363E-02	4.865E-03	4.344E-02	3.107E-01	3.919E-02	4.009E-02	8.543E-03
14	1.258E-02	7.502E-02	1.655E-03	1.761E-03	2.491E-03	9.004E-04	2.603E-04	5.659E-03	1.344E-03	2.904E-03	2.630E-03	2.312E-03	1.453E-03	3.467E-01	2.595E-03	4.444E-03
15	1.020E-04	1.359E-04	7.497E-05	5.684E-05	8.890E-05	9.749E-05	2.975E-05	1.316E-04	5.084E-05	3.234E-04	1.101E-04	2.927E-03	1.148E-03	5.194E-04	1.265E-01	2.979E-04
16	9.786E-03	1.369E-02	2.324E-02	3.621E-02	2.319E-02	2.403E-02	1.414E-03	1.202E-02	3.337E-03	1.700E-02	1.256E-02	2.878E-02	4.324E-02	4.437E-02	6.477E-02	1.544E-01
17	2.430E-03	4.515E-03	1.816E-03	3.022E-03	4.413E-03	4.507E-03	1.734E-03	9.764E-03	5.777E-03	4.464E-03	5.414E-03	2.130E-03	3.073E-03	1.637E-03	3.516E-03	2.332E-03
18	3.805E-03	2.285E-02	1.281E-02	1.796E-02	2.471E-02	3.648E-02	1.058E-02	3.381E-02	4.055E-02	3.361E-02	1.904E-02	1.286E-02	1.393E-02	9.851E-03	1.194E-02	1.700E-02
19	9.682E-04	2.751E-03	3.613E-03	3.821E-03	3.205E-03	5.783E-03	9.781E-04	4.029E-03	2.309E-03	2.079E-03	1.562E-03	2.446E-03	2.437E-03	1.355E-03	2.817E-03	1.484E-03
20	4.055E-02	4.586E-02	6.997E-02	9.448E-02	7.555E-02	2.961E-02	1.558E-02	4.268E-02	4.134E-02	5.050E-02	3.690E-02	5.806E-02	5.440E-02	5.062E-02	5.902E-02	4.634E-02
21	4.524E-02	7.341E-02	1.031E-02	4.571E-02	2.451E-02	2.424E-02	2.092E-02	3.114E-02	1.661E-02	2.032E-02	2.105E-02	1.645E-02	9.438E-03	1.178E-02	3.005E-02	1.788E-02
22	4.232E-04	9.366E-03	2.709E-03	4.817E-03	5.605E-03	9.275E-03	2.239E-03	9.270E-03	4.032E-03	4.158E-03	6.876E-03	6.078E-03	8.190E-03	4.868E-03	9.225E-03	6.959E-03
23	2.665E-02	4.180E-02	3.055E-02	2.021E-02	2.983E-02	2.447E-02	1.957E-02	4.921E-02	1.774E-02	1.874E-02	2.694E-02	1.548E-02	1.465E-02	1.222E-02	1.466E-02	2.163E-02
24	1.030E-03	4.232E-03	3.073E-03	4.454E-03	2.881E-03	7.777E-03	1.839E-03	2.944E-03	1.283E-03	3.645E-03	5.865E-03	3.836E-03	4.874E-03	4.141E-03	5.310E-03	6.300E-03
25	2.518E-04	5.052E-04	7.300E-04	1.016E-03	4.541E-04	1.345E-03	1.195E-04	1.058E-03	3.703E-04	5.039E-04	4.271E-04	2.837E-04	2.932E-04	7.444E-04	6.298E-04	7.185E-04
26	4.547E-06	7.101E-05	4.782E-04	5.647E-04	3.741E-04	6.303E-03	3.651E-05	5.535E-04	3.218E-04	1.249E-03	8.643E-04	1.422E-03	2.333E-03	1.605E-03	2.531E-03	1.110E-03
27	2.671E-04	1.293E-03	6.630E-04	4.283E-04	7.168E-04	8.969E-04	2.674E-04	8.197E-04	3.525E-04	4.919E-04	1.041E-03	6.339E-04	7.932E-04	4.801E-04	6.572E-04	9.222E-04
28	1.785E-04	9.393E-04	8.205E-04	1.068E-03	9.102E-04	1.566E-03	4.571E-04	1.054E-03	6.643E-04	4.705E-04	1.735E-03	1.464E-03	7.044E-04	6.161E-04	6.459E-04	9.670E-04
29	5.276E-03	4.210E-02	3.082E-02	2.234E-02	2.667E-02	5.688E-02	5.706E-03	2.399E-02	1.302E-02	2.035E-02	3.505E-02	2.983E-02	4.015E-02	2.438E-02	3.806E-02	4.316E-02
30	6.980E-03	3.194E-02	1.679E-02	1.098E-02	1.810E-02	2.237E-02	6.655E-03	2.056E-02	8.923E-03	1.240E-02	2.605E-02	1.608E-02	2.076E-02	1.227E-02	1.692E-02	2.369E-02
31	5.829E-05	8.977E-05	1.118E-03	6.409E-04	1.830E-03	2.361E-03	3.036E-04	2.458E-03	4.193E-03	2.595E-03	1.009E-03	7.963E-04	2.429E-03	1.811E-03	8.312E-04	2.548E-03
32	1.687E-03	7.717E-03	1.671E-03	3.517E-03	6.902E-03	6.159E-03	4.632E-04	1.305E-02	4.080E-03	4.829E-03	6.652E-03	3.281E-03	4.396E-03	1.794E-03	3.548E-03	4.293E-03
v1	5.546E-01	5.013E-01	3.604E-01	3.654E-01	3.489E-01	3.487E-01	4.329E-01	4.284E-01	2.824E-01	3.192E-01	4.435E-01	3.913E-01	3.312E-01	2.693E-01	4.550E-01	4.211E-01

Table 7.17 Input coefficients of usual industries and other parameters (1997) (no.2)

Industry	17	18	19	20	21	22	23	24	25	26	27	28	29	30	31	32
1	1.61E-03	2.15E-04	2.14E-04	2.77E-03	2.84E-04	4.25E-05	2.73E-04	2.77E-04	1.08E-03	2.49E-04	4.78E-03	2.40E-03	2.97E-04	1.49E-02	1.14E-04	2.12E-03
2	9.04E-03	8.99E-02	1.19E-05	2.40E-07	0.00E+00	0.00E+00	4.58E-06	0.00E+00	3.38E-05	1.49E-04	3.25E-06	1.33E-06	3.32E-06	1.54E-05	0.00E+00	8.69E-04
3	1.67E-03	2.05E-03	2.57E-03	2.74E-03	3.14E-03	6.81E-04	2.42E-03	2.61E-03	3.14E-03	2.36E-02	1.70E-02	7.09E-03	3.77E-03	1.07E-01	1.51E-03	2.03E-03
4	1.68E-03	3.77E-04	1.48E-03	2.71E-03	9.13E-04	9.59E-05	1.92E-03	9.46E-04	3.44E-03	1.88E-04	3.18E-03	1.07E-02	2.41E-03	1.89E-03	6.91E-03	2.13E-02
5	4.42E-02	1.48E-03	2.07E-03	1.33E-02	5.30E-03	9.35E-04	3.69E-03	2.76E-03	3.95E-03	4.18E-03	3.93E-03	1.66E-02	1.15E-02	5.77E-03	2.32E-01	3.18E-02
6	4.52E-03	1.35E-03	8.55E-03	7.71E-04	9.75E-04	1.41E-04	1.04E-03	2.31E-03	4.13E-03	7.50E-04	1.86E-01	7.42E-03	3.38E-03	5.55E-03	2.22E-03	3.71E-02
7	1.30E-02	3.83E-02	4.86E-03	8.73E-03	1.07E-03	1.11E-03	3.22E-02	1.98E-03	1.07E-02	4.68E-03	6.61E-03	4.77E-03	5.52E-03	6.06E-03	2.13E-03	2.46E-02
8	5.69E-02	1.51E-04	2.56E-03	8.79E-04	1.77E-04	5.65E-06	2.45E-04	1.36E-04	8.08E-04	1.38E-03	1.47E-03	1.19E-03	2.56E-04	2.80E-03	7.95E-05	9.13E-03
9	1.91E-02	0.00E+00	4.16E-06	2.23E-04	1.08E-07	0.00E+00	2.32E-04	0.00E+00	1.08E-04	0.00E+00	8.19E-06	1.92E-06	6.11E-06	2.12E-05	2.25E-02	1.33E-02
10	6.30E-03	4.88E-04	2.20E-04	7.34E-05	6.46E-06	1.18E-06	7.45E-05	5.17E-06	1.64E-04	3.61E-07	1.74E-03	1.31E-04	2.72E-05	2.85E-04	2.15E-04	1.17E-02
11	7.97E-02	5.51E-04	8.59E-04	3.88E-03	1.62E-04	3.73E-05	1.22E-03	3.46E-04	5.18E-03	9.33E-05	3.81E-04	1.26E-03	5.50E-04	1.85E-03	1.53E-02	1.96E-02
12	1.51E-02	1.82E-03	7.46E-03	1.53E-03	2.68E-03	5.88E-04	2.58E-03	7.28E-04	7.06E-03	3.84E-03	6.58E-04	7.22E-03	8.45E-03	4.18E-03	4.66E-03	1.48E-03
13	1.24E-02	6.17E-02	8.59E-03	5.90E-04	9.65E-04	2.40E-04	1.21E-03	4.57E-03	1.13E-02	2.91E-03	9.76E-04	4.14E-04	1.32E-02	3.23E-03	7.32E-04	3.41E-02
14	7.26E-03	2.16E-03	3.25E-03	9.65E-03	7.90E-04	5.63E-04	2.94E-02	1.66E-03	3.12E-02	2.18E-03	2.18E-03	2.27E-03	4.29E-03	3.89E-03	1.60E-03	3.88E-03
15	8.51E-05	6.00E-04	1.22E-03	1.77E-03	8.93E-05	2.56E-05	1.05E-04	8.23E-05	1.09E-03	3.83E-04	8.54E-03	8.82E-04	6.41E-04	4.61E-04	1.96E-04	3.66E-04
16	1.92E-02	5.79E-03	3.06E-02	1.35E-02	2.32E-02	9.57E-04	6.97E-03	2.33E-02	4.36E-02	1.81E-02	9.69E-03	5.55E-02	6.02E-02	1.25E-02	2.65E-02	4.72E-02
17	2.48E-03	2.42E-02	3.31E-02	3.71E-03	2.72E-03	3.28E-02	8.41E-03	1.57E-03	1.81E-02	1.14E-02	2.65E-03	4.68E-03	3.23E-03	5.11E-03	1.74E-03	1.23E-03
18	6.24E-03	4.39E-03	5.80E-02	8.16E-03	3.09E-03	3.61E-03	1.51E-02	1.02E-02	1.89E-02	9.40E-03	2.07E-02	7.63E-03	7.01E-03	1.61E-02	2.45E-03	2.10E-02
19	1.87E-03	5.62E-03	3.12E-03	1.13E-03	1.38E-03	8.03E-04	5.01E-03	1.98E-03	1.54E-02	7.50E-03	8.19E-03	3.95E-03	2.56E-03	1.31E-02	7.55E-04	8.25E-03
20	5.48E-02	1.19E-02	1.14E-02	1.94E-02	5.39E-03	1.67E-03	1.52E-02	7.09E-03	1.73E-02	9.62E-03	6.40E-02	2.63E-02	2.14E-02	4.92E-02	4.77E-02	5.21E-02
21	1.58E-02	3.05E-02	1.56E-02	4.57E-02	1.61E-01	5.41E-02	6.81E-02	1.02E-02	5.96E-03	1.88E-03	9.71E-03	1.91E-02	2.75E-02	1.72E-02	8.17E-03	3.02E-02
22	4.35E-03	1.38E-02	4.40E-03	5.06E-02	3.51E-02	8.25E-03	2.10E-02	1.07E-02	2.91E-03	3.41E-03	4.86E-03	1.79E-02	2.43E-02	2.40E-02	9.76E-03	1.72E-02
23	2.31E-02	1.33E-02	1.50E-02	2.30E-02	1.40E-02	1.04E-03	8.28E-02	1.47E-02	2.26E-02	7.74E-03	1.22E-02	1.38E-02	1.60E-02	1.53E-02	2.33E-02	1.95E-02
24	6.78E-03	3.87E-03	5.96E-03	2.51E-02	2.27E-02	6.30E-04	1.05E-02	6.61E-02	1.38E-02	4.51E-03	5.32E-03	1.70E-02	6.53E-02	9.01E-03	1.35E-02	9.25E-03
25	6.35E-04	4.57E-04	1.41E-03	1.88E-03	4.24E-04	3.10E-04	1.51E-02	8.01E-04	1.44E-03	8.58E-04	3.73E-03	5.50E-04	6.33E-04	2.72E-03	3.83E-04	8.15E-02
26	5.69E-04	4.00E-03	1.36E-04	4.82E-04	3.35E-04	2.39E-06	1.09E-03	2.63E-02	2.18E-04	2.70E-07	1.66E-04	2.23E-05	1.12E-03	1.40E-04	0.00E+00	3.90E-03
27	5.29E-04	6.40E-04	8.03E-04	6.93E-04	8.63E-04	1.88E-04	9.51E-04	9.55E-04	2.64E-06	1.01E-04	1.18E-03	1.34E-04	1.27E-03	8.07E-04	5.19E-04	6.37E-04
28	9.56E-04	1.14E-03	9.83E-03	6.20E-04	2.61E-03	2.04E-04	1.40E-03	1.58E-03	6.12E-05	4.73E-04	1.19E-03	1.37E-05	2.57E-03	2.98E-03	2.52E-05	1.92E-04
29	7.88E-02	6.86E-02	5.57E-02	8.74E-02	1.49E-01	1.91E-02	5.32E-02	1.41E-01	9.33E-02	2.94E-02	4.38E-02	6.69E-02	1.53E-01	4.10E-02	3.13E-02	8.21E-02
30	1.36E-02	1.65E-02	1.93E-02	1.83E-02	2.17E-02	5.41E-03	2.30E-02	6.09E-02	2.02E-03	3.03E-03	2.16E-02	7.49E-03	3.72E-02	3.78E-02	1.36E-02	2.91E-02
31	1.72E-05	3.43E-06	1.70E-06	8.05E-03	4.55E-06	1.02E-06	3.30E-03	4.54E-06	1.50E-04	3.16E-07	1.13E-06	0.00E+00	4.68E-06	7.07E-07	0.00E+00	1.38E-03
32	8.70E-03	4.34E-03	5.66E-03	2.91E-03	4.36E-03	7.94E-03	3.83E-03	2.56E-03	2.83E-02	8.59E-03	8.18E-03	4.88E-02	4.36E-03	6.46E-03	1.04E-02	2.62E-04
v1	0.48906	0.589778	0.686162	0.639841	0.535504	0.858449	0.588316	0.601753	0.632675	0.839388	0.545636	0.647834	0.518271	0.58828	0.520033	0.381741

Table 7.18 Input coefficients of usual industries and other parameters (2000) (no.1)

Industry	1	2	3	4	5	6	7	8	9	10	11	12	13	14	15	16
1	1.085E-01	3.794E-04	1.942E-01	8.142E-03	3.899E-02	2.121E-03	2.865E-05	5.210E-05	8.159E-07	2.444E-05	0.000E+00	0.000E+00	0.000E+00	2.578E-06	0.000E+00	5.373E-03
2	1.454E-05	2.531E-03	7.964E-07	3.806E-06	1.336E-03	2.521E-03	4.447E-01	6.904E-02	2.364E-02	6.265E-02	1.047E-04	3.211E-05	3.277E-06	1.652E-05	2.412E-05	3.885E-03
3	7.448E-02	0.000E+00	1.383E-01	2.235E-03	1.665E-03	4.127E-03	4.621E-06	4.575E-04	1.107E-06	0.000E+00	0.000E+00	0.000E+00	0.000E+00	0.000E+00	0.000E+00	1.465E-03
4	4.707E-03	5.606E-03	1.264E-03	2.622E-01	6.223E-03	8.641E-04	3.904E-04	2.751E-03	6.454E-04	1.884E-03	1.390E-03	1.352E-03	2.558E-03	2.023E-03	1.554E-03	4.394E-03
5	1.475E-02	2.780E-03	1.719E-02	9.720E-03	2.591E-01	1.596E-02	4.721E-05	2.050E-02	9.190E-04	5.502E-03	4.910E-03	2.003E-03	6.166E-03	1.668E-03	6.546E-03	5.608E-02
6	4.550E-02	8.967E-03	8.944E-03	8.457E-02	3.319E-02	2.932E-01	2.970E-03	2.415E-02	6.085E-03	1.491E-02	9.547E-03	6.115E-03	9.960E-03	1.027E-02	7.019E-03	1.014E-01
7	1.309E-02	1.010E-02	2.857E-03	3.657E-03	7.991E-03	3.924E-02	5.064E-02	1.236E-02	1.530E-02	5.048E-03	2.380E-03	1.241E-03	8.269E-04	1.391E-03	7.106E-04	1.737E-02
8	1.173E-03	1.371E-04	4.325E-03	5.699E-04	6.367E-03	6.357E-03	6.468E-04	8.668E-02	8.279E-03	7.796E-03	3.827E-03	5.531E-03	1.176E-02	7.127E-03	1.845E-02	3.255E-03
9	9.423E-05	1.206E-03	0.000E+00	3.651E-05	7.433E-03	4.222E-05	0.000E+00	7.665E-03	4.553E-01	1.505E-03	1.734E-01	6.434E-02	1.280E-02	4.065E-02	1.021E-02	7.147E-03
10	0.000E+00	1.625E-04	1.226E-03	1.889E-05	1.765E-03	3.858E-03	1.101E-05	4.036E-03	7.238E-03	2.963E-01	5.670E-02	1.994E-02	3.028E-02	1.736E-02	2.140E-02	8.675E-03
11	1.168E-03	1.913E-02	1.907E-02	2.177E-03	1.307E-02	9.864E-03	1.422E-03	9.328E-03	9.731E-04	2.188E-03	5.993E-02	3.371E-02	1.577E-02	9.902E-03	1.843E-02	7.429E-03
12	1.907E-05	4.845E-03	3.340E-07	0.000E+00	1.870E-03	3.195E-05	8.395E-06	2.812E-03	4.905E-04	4.958E-04	2.015E-03	1.965E-01	6.921E-03	1.174E-02	9.422E-03	1.712E-03
13	2.615E-04	4.591E-04	1.220E-05	1.819E-05	1.918E-04	7.685E-05	5.854E-06	1.673E-05	9.208E-06	3.819E-04	6.002E-03	5.819E-02	2.990E-01	4.453E-02	8.278E-02	7.151E-03
14	5.143E-03	3.989E-05	0.000E+00	0.000E+00	0.000E+00	0.000E+00	0.000E+00	0.000E+00	0.000E+00	0.000E+00	0.000E+00	0.000E+00	0.000E+00	4.315E-01	0.000E+00	0.000E+00
15	7.439E-05	2.031E-05	2.980E-06	1.156E-05	4.199E-05	2.364E-05	9.243E-07	4.075E-05	2.972E-06	5.539E-06	3.211E-05	4.461E-03	8.476E-04	6.468E-04	1.131E-01	1.027E-04
16	1.051E-02	1.276E-02	2.747E-02	4.049E-02	3.052E-02	2.515E-02	4.267E-03	2.558E-02	3.098E-02	6.084E-02	1.241E-02	2.497E-02	3.987E-02	4.135E-02	4.709E-02	1.651E-01
17	5.630E-03	6.585E-03	1.920E-03	4.127E-03	6.270E-03	6.950E-03	2.001E-03	1.476E-02	8.442E-03	6.318E-03	9.455E-03	2.830E-03	3.565E-03	1.455E-03	4.164E-03	3.016E-03
18	5.675E-03	2.728E-02	1.067E-02	1.396E-02	3.344E-02	3.962E-02	9.405E-03	3.708E-02	3.997E-02	3.674E-02	1.893E-02	9.963E-03	1.223E-02	9.356E-03	1.265E-02	1.687E-02
19	7.219E-04	3.995E-03	3.192E-03	3.554E-03	3.334E-03	7.361E-03	9.206E-04	5.084E-03	2.614E-03	1.969E-03	1.524E-03	2.154E-03	1.382E-03	1.065E-03	2.230E-03	1.347E-03
20	4.634E-02	1.687E-02	7.755E-02	6.631E-02	6.885E-02	3.923E-02	1.185E-02	4.584E-02	3.395E-02	4.305E-02	4.558E-02	5.541E-02	5.389E-02	4.188E-02	6.347E-02	5.373E-02
21	3.497E-02	4.789E-02	6.800E-03	3.033E-02	1.794E-02	1.663E-02	1.180E-02	2.922E-02	1.377E-02	1.956E-02	1.757E-02	1.381E-02	9.088E-03	8.708E-03	2.179E-02	1.564E-02
22	4.246E-04	8.925E-03	1.868E-03	5.544E-03	4.104E-03	3.752E-03	9.509E-04	4.469E-03	2.925E-03	2.161E-03	5.225E-03	3.536E-03	2.446E-03	1.145E-03	4.038E-03	4.345E-03
23	4.310E-02	2.752E-01	3.234E-02	2.546E-02	3.869E-02	2.519E-02	2.959E-02	7.284E-02	2.910E-02	2.759E-02	2.991E-02	1.918E-02	1.662E-02	1.529E-02	1.779E-02	3.833E-02
24	8.371E-04	6.518E-03	1.491E-03	5.283E-03	2.712E-03	7.032E-03	1.453E-03	3.042E-03	1.319E-03	4.196E-03	6.476E-03	4.421E-03	3.712E-03	1.749E-03	4.611E-03	6.185E-03
25	0.000E+00	0.000E+00	0.000E+00	0.000E+00	0.000E+00	0.000E+00	0.000E+00	0.000E+00	0.000E+00	0.000E+00	0.000E+00	0.000E+00	0.000E+00	0.000E+00	0.000E+00	0.000E+00
26	8.755E-04	2.130E-03	5.653E-03	6.538E-03	6.045E-03	7.720E-02	4.119E-03	3.339E-02	1.005E-02	2.978E-02	9.956E-03	3.108E-02	7.481E-02	3.346E-02	6.691E-02	1.257E-02
27	0.000E+00	0.000E+00	0.000E+00	0.000E+00	1.346E-06	1.172E-05	0.000E+00	0.000E+00	2.448E-06	0.000E+00	0.000E+00	0.000E+00	0.000E+00	0.000E+00	0.000E+00	1.440E-06
28	2.478E-04	1.516E-03	7.352E-04	1.073E-03	7.567E-04	1.541E-03	3.470E-04	1.072E-03	8.302E-04	3.539E-04	1.199E-03	1.653E-03	8.020E-04	2.981E-04	6.644E-04	8.163E-04
29	1.346E-02	4.134E-02	3.651E-02	3.572E-02	3.309E-02	6.147E-02	9.198E-03	4.698E-02	2.330E-02	2.859E-02	4.295E-02	4.469E-02	5.281E-02	2.893E-02	4.524E-02	4.818E-02
30	2.860E-04	2.713E-04	1.969E-04	3.156E-04	2.473E-04	2.474E-04	7.425E-05	2.678E-04	2.441E-04	2.361E-04	2.062E-04	2.467E-04	2.683E-04	1.855E-04	2.125E-04	1.049E-03
31	4.244E-04	1.293E-03	1.179E-03	2.156E-03	1.239E-03	6.366E-04	7.094E-05	1.305E-03	4.519E-04	6.533E-04	1.803E-03	1.496E-03	1.485E-03	5.864E-04	1.148E-03	1.353E-03
32	5.645E-03	1.483E-02	5.528E-03	1.000E-02	6.178E-03	6.027E-03	1.835E-03	4.559E-03	8.380E-03	6.210E-03	1.156E-02	1.003E-02	3.519E-03	1.572E-03	4.582E-03	5.328E-03
v1	5.619E-01	4.762E-01	3.995E-01	3.758E-01	3.674E-01	3.037E-01	4.113E-01	4.346E-01	2.748E-01	3.331E-01	4.650E-01	3.811E-01	3.266E-01	2.341E-01	4.138E-01	4.007E-01

Table 7.19 Input coefficients of usual industries and other parameters (2000) (no.2)

Industry	17	18	19	20	21	22	23	24	25	26	27	28	29	30	31	32
1	1.967E-03	0.000E+00	0.000E+00	9.259E-05	0.000E+00	1.397E-06	4.386E-05	0.000E+00	5.775E-05	6.976E-04	4.329E-03	2.160E-03	1.457E-05	1.892E-02	0.000E+00	0.000E+00
2	8.717E-03	1.047E-01	1.089E-05	0.000E+00	0.000E+00	0.000E+00	8.141E-07	0.000E+00	1.711E-05	9.602E-05	3.613E-06	0.000E+00	9.705E-07	4.329E-06	0.000E+00	2.345E-04
3	0.000E+00	0.000E+00	0.000E+00	1.577E-04	0.000E+00	0.000E+00	1.856E-04	0.000E+00	2.550E-04	4.461E-04	1.343E-02	1.576E-03	0.000E+00	1.022E-01	0.000E+00	2.388E-04
4	2.583E-03	2.654E-04	1.318E-03	2.944E-03	1.344E-03	2.466E-05	1.875E-03	7.093E-04	2.042E-03	1.995E-04	3.348E-03	1.607E-02	1.702E-03	3.502E-03	1.709E-02	7.219E-03
5	4.825E-02	1.071E-03	3.219E-03	7.549E-03	4.361E-03	5.986E-04	6.499E-03	2.174E-03	1.928E-03	4.298E-03	4.694E-03	1.452E-02	4.313E-03	6.286E-03	4.223E-01	1.609E-02
6	4.809E-03	1.015E-03	1.367E-02	6.385E-06	2.052E-05	1.960E-05	4.135E-04	7.357E-04	7.961E-04	4.302E-03	1.341E-01	1.844E-03	3.162E-03	6.981E-03	5.785E-02	1.673E-02
7	9.486E-03	3.339E-02	8.528E-03	1.252E-03	3.409E-04	5.618E-04	9.733E-02	9.140E-04	5.111E-03	4.886E-03	3.227E-03	3.674E-03	1.227E-03	3.311E-03	0.000E+00	7.537E-03
8	6.161E-02	6.097E-05	2.515E-03	3.477E-04	1.738E-05	4.164E-05	6.945E-05	4.517E-08	2.124E-04	1.478E-03	1.229E-03	1.085E-03	7.283E-04	2.639E-03	2.910E-03	6.940E-03
9	1.674E-02	0.000E+00	3.971E-04	0.000E+00	0.000E+00	0.000E+00	3.059E-04	0.000E+00	1.996E-05	0.000E+00	7.249E-06	4.489E-06	8.583E-05	2.368E-05	1.466E-05	5.623E-03
10	7.315E-03	5.183E-04	1.155E-04	1.076E-05	0.000E+00	0.000E+00	1.108E-05	0.000E+00	1.443E-04	0.000E+00	1.017E-03	1.238E-04	1.055E-04	3.205E-04	5.282E-04	3.084E-03
11	8.994E-02	7.914E-04	6.096E-04	2.837E-03	5.963E-05	2.506E-04	1.492E-03	2.318E-04	4.686E-03	7.200E-05	3.038E-04	1.692E-03	1.122E-03	2.128E-03	1.688E-04	5.288E-03
12	6.577E-03	6.792E-06	3.615E-03	6.313E-06	0.000E+00	0.000E+00	8.961E-05	9.621E-06	3.128E-04	0.000E+00	0.000E+00	0.000E+00	1.998E-02	2.475E-04	4.117E-02	0.000E+00
13	1.082E-02	2.888E-05	1.151E-04	3.667E-04	1.294E-04	2.323E-05	2.974E-04	1.217E-03	8.916E-03	1.533E-03	7.844E-05	6.828E-05	1.482E-02	7.938E-04	1.028E-02	2.809E-03
14	0.000E+00	0.000E+00	0.000E+00	0.000E+00	0.000E+00	0.000E+00	1.107E-02	0.000E+00	2.303E-02	2.838E-05	0.000E+00	0.000E+00	2.327E-02	2.929E-05	0.000E+00	0.000E+00
15	8.496E-05	0.000E+00	8.359E-05	1.455E-03	4.986E-05	3.720E-06	2.747E-05	3.090E-05	6.760E-04	1.105E-05	6.349E-03	1.654E-05	5.970E-04	3.158E-04	0.000E+00	0.000E+00
16	2.049E-02	2.168E-02	3.138E-02	1.223E-02	2.275E-02	5.930E-04	5.737E-03	1.266E-02	2.941E-02	3.176E-02	9.620E-03	7.168E-02	4.950E-02	1.356E-02	1.419E-01	1.929E-02
17	2.574E-03	5.389E-02	2.842E-02	5.655E-03	3.939E-03	4.318E-02	1.000E-02	7.856E-03	1.571E-02	1.385E-02	5.550E-03	2.008E-03	2.826E-03	7.009E-03	0.000E+00	0.000E+00
18	4.715E-03	3.465E-02	5.120E-02	9.920E-03	3.589E-03	2.856E-03	1.383E-02	9.189E-03	1.255E-02	2.364E-02	1.741E-02	4.409E-03	6.094E-03	2.386E-02	0.000E+00	4.111E-03
19	2.260E-03	5.751E-03	5.817E-02	2.728E-03	2.329E-03	4.762E-04	4.709E-03	5.190E-03	1.607E-02	9.937E-03	9.256E-03	3.112E-03	8.927E-04	1.941E-02	0.000E+00	1.172E-02
20	6.394E-02	1.383E-02	1.621E-02	1.458E-02	5.004E-03	9.519E-04	3.434E-02	4.324E-03	1.245E-02	1.426E-02	5.414E-02	2.944E-02	2.316E-02	6.083E-02	2.553E-01	1.998E-02
21	1.118E-02	3.486E-02	1.151E-02	5.083E-02	7.528E-02	5.009E-02	6.122E-02	2.274E-02	2.763E-03	7.152E-03	1.387E-02	1.458E-02	4.619E-02	2.364E-02	0.000E+00	2.269E-01
22	3.481E-03	1.050E-02	2.792E-03	2.952E-02	1.597E-02	6.193E-03	1.469E-02	1.699E-02	1.288E-03	9.722E-03	7.330E-03	2.518E-02	1.203E-02	1.659E-02	0.000E+00	1.171E-02
23	5.159E-02	2.310E-02	3.501E-02	4.787E-02	1.909E-02	2.224E-03	1.051E-01	2.278E-02	3.100E-02	1.494E-02	1.619E-02	2.719E-02	1.579E-02	2.781E-02	5.038E-02	4.941E-02
24	1.214E-02	4.057E-03	8.363E-03	2.599E-02	2.174E-02	1.492E-03	7.527E-03	1.207E-01	1.460E-02	1.196E-02	6.402E-03	2.808E-02	4.279E-02	1.290E-02	0.000E+00	2.800E-02
25	0.000E+00	0.000E+00	0.000E+00	0.000E+00	0.000E+00	0.000E+00	0.000E+00	0.000E+00	0.000E+00	0.000E+00	0.000E+00	0.000E+00	0.000E+00	0.000E+00	0.000E+00	1.683E-01
26	2.674E-03	1.873E-02	1.501E-04	1.571E-03	4.861E-04	1.944E-06	1.815E-03	1.487E-02	1.991E-04	4.684E-07	1.508E-04	0.000E+00	2.662E-03	3.213E-04	0.000E+00	1.319E-02
27	5.174E-07	1.866E-06	2.423E-05	1.447E-05	2.621E-05	2.566E-06	2.451E-05	4.842E-05	7.674E-06	5.015E-06	1.813E-02	9.924E-06	5.522E-06	4.185E-05	0.000E+00	0.000E+00
28	9.341E-04	1.437E-03	6.317E-03	4.795E-04	2.156E-03	2.382E-04	1.035E-03	9.005E-04	6.404E-06	1.184E-03	1.106E-03	0.000E+00	1.837E-03	3.362E-03	0.000E+00	1.135E-03
29	7.864E-02	9.018E-02	7.292E-02	6.168E-02	1.276E-01	2.493E-02	1.359E-01	1.127E-01	7.415E-02	4.820E-02	4.706E-02	8.352E-02	1.126E-01	3.848E-02	0.000E+00	5.651E-02
30	5.106E-04	4.200E-04	6.505E-04	1.705E-03	6.903E-04	7.041E-04	7.472E-04	3.596E-02	1.788E-03	1.196E-03	1.585E-02	5.577E-03	8.198E-03	2.416E-02	0.000E+00	1.223E-02
31	4.188E-04	1.187E-03	2.170E-03	3.765E-03	4.029E-03	3.463E-04	1.798E-03	2.181E-03	2.417E-03	4.089E-03	2.452E-03	5.214E-03	3.226E-03	2.285E-03	0.000E+00	2.609E-04
32	3.981E-03	4.870E-03	6.478E-03	6.466E-03	6.741E-03	4.104E-03	3.985E-03	4.589E-03	4.952E-04	1.106E-03	1.917E-03	6.690E-03	4.396E-03	3.934E-03	1.710E-04	0.000E+00
v1	4.716E-01	5.390E-01	6.340E-01	7.080E-01	6.822E-01	8.601E-01	4.779E-01	6.003E-01	7.369E-01	7.890E-01	6.014E-01	6.505E-01	5.967E-01	5.742E-01	0.000E+00	3.055E-01

Table 7.20 Input coefficients of abatement industries and other parameters (1997)

	By abatement activity of		
Demand for industry of	CO_2 (million yen/t)	SOx (million yen/t)	NOx (million yen/t)
1	2.043E-05	8.073E-05	5.848E-04
2	2.070E-05	8.181E-05	5.926E-04
3	1.675E-04	6.618E-04	4.794E-03
4	7.129E-05	2.818E-04	2.041E-03
5	2.562E-04	1.012E-03	7.334E-03
6	5.112E-04	2.020E-03	1.464E-02
7	1.990E-04	7.864E-04	5.697E-03
8	3.656E-04	1.445E-03	1.047E-02
9	5.683E-03	2.246E-02	1.627E-01
10	1.689E-03	6.678E-03	4.837E-02
11	2.688E-03	1.062E-02	7.696E-02
12	1.858E-02	7.343E-02	5.319E-01
13	3.649E-03	1.442E-02	1.045E-01
14	1.942E-04	7.674E-04	5.559E-03
15	2.459E-04	9.718E-04	7.040E-03
16	2.418E-03	9.555E-03	6.922E-02
17	1.789E-04	7.072E-04	5.123E-03
18	1.080E-03	4.269E-03	3.093E-02
19	2.055E-04	8.120E-04	5.882E-03
20	4.877E-03	1.927E-02	1.396E-01
21	1.382E-03	5.461E-03	3.956E-02
22	5.105E-04	2.018E-03	1.462E-02
23	1.301E-03	5.141E-03	3.724E-02
24	3.223E-04	1.274E-03	9.227E-03
25	2.383E-05	9.420E-05	6.824E-04
26	1.194E-04	4.721E-04	3.420E-03
27	5.325E-05	2.104E-04	1.524E-03
28	1.230E-04	4.861E-04	3.521E-03
29	2.506E-03	9.903E-03	7.174E-02
30	1.351E-03	5.338E-03	3.867E-02
31	6.689E-05	2.644E-04	1.915E-03
32	2.756E-04	1.089E-03	7.890E-03
v2	3.287E-02	1.299E-01	9.410E-01
Total	0.084	0.332	2.405

Table 7.21 Input coefficients of abatement industries and other parameters (2000)

	By abatement activity of		
Demand for industry of	CO_2 (million yen/t)	SOx (million yen/t)	NOx (million yen/t)
1	0.000E+00	0.000E+00	0.000E+00
2	2.697E-06	1.066E-05	7.723E-05
3	0.000E+00	0.000E+00	0.000E+00
4	1.135E-04	4.487E-04	3.251E-03
5	1.683E-04	6.651E-04	4.818E-03
6	5.137E-04	2.030E-03	1.471E-02
7	1.042E-04	4.120E-04	2.984E-03
8	4.646E-04	1.836E-03	1.330E-02
9	5.404E-03	2.136E-02	1.547E-01
10	1.675E-03	6.619E-03	4.795E-02
11	2.832E-03	1.119E-02	8.108E-02
12	1.650E-02	6.523E-02	4.725E-01

(continued)

Table 7.21 (continued)

Demand for industry of	By abatement activity of		
	CO_2 (million yen/t)	SOx (million yen/t)	NOx (million yen/t)
13	4.888E-03	1.932E-02	1.400E-01
14	0.000E+00	0.000E+00	0.000E+00
15	3.748E-04	1.481E-03	1.073E-02
16	2.098E-03	8.291E-03	6.006E-02
17	2.377E-04	9.396E-04	6.807E-03
18	8.369E-04	3.308E-03	2.396E-02
19	1.810E-04	7.153E-04	5.181E-03
20	4.654E-03	1.839E-02	1.333E-01
21	1.160E-03	4.584E-03	3.320E-02
22	2.970E-04	1.174E-03	8.504E-03
23	1.611E-03	6.367E-03	4.612E-02
24	3.714E-04	1.468E-03	1.063E-02
25	0.000E+00	0.000E+00	0.000E+00
26	2.611E-03	1.032E-02	7.475E-02
27	0.000E+00	0.000E+00	0.000E+00
28	1.389E-04	5.489E-04	3.976E-03
29	3.754E-03	1.484E-02	1.075E-01
30	2.072E-05	8.191E-05	5.934E-04
31	1.256E-04	4.966E-04	3.597E-03
32	8.426E-04	3.330E-03	2.412E-02
v2	3.202E-02	1.265E-01	9.166E-01
Total	0.084	0.332	2.405

Table 7.22 Emission coefficients in Japan (1997)

Industry	Pollutants		
	CO_2 (t/million yen)	SOx (t/million yen)	NOx (t/million yen)
1 Agriculture	1.21E+00	3.85E-03	2.05E-02
2 Mining	4.65E-01	2.99E-04	7.45E-04
3 Food products	3.79E-01	1.93E-03	4.15E-04
4 Textiles and clothing	3.94E-01	1.71E-03	5.15E-04
5 Paper and wooden products	1.75E+00	4.96E-03	1.82E-03
6 Chemical products	1.82E+00	3.92E-03	2.81E-03
7 Coal and petroleum products	6.45E+00	6.23E-03	5.90E-03
8 Nonmetallic mineral products	9.80E+00	4.49E-03	1.57E-02
9 Iron and steel	3.85E+00	3.77E-03	3.70E-03
10 Nonferrous metals	1.04E+00	2.64E-03	2.81E-03
11 Fabricated metal products	2.21E-01	1.46E-04	2.08E-04
12 General machinery	1.23E-01	1.61E-04	1.24E-04
13 Electric machinery	7.42E-02	8.68E-05	7.35E-05
14 Transport equipment, automobiles	1.25E-01	2.14E-04	1.63E-04
15 Precision apparatus	7.99E-02	1.22E-04	7.62E-05
16 Other manufactures	1.74E-01	6.34E-04	2.35E-04
17 Building and construction	1.60E-01	9.09E-05	1.27E-04
18 Electricity, gas	2.24E+01	1.79E-02	1.67E-02
19 Water	6.07E+00	6.12E-03	7.00E-03
20 Wholesale and retail service	1.82E-01	3.50E-04	1.35E-04
21 Finance and insurance	2.42E-02	9.19E-06	1.72E-05
22 Real estate services	6.35E-02	5.19E-06	6.28E-05
23 Transport	5.10E+00	1.92E-02	6.33E-02
24 Communications	8.41E-02	9.73E-05	4.59E-05

(continued)

Table 7.22 (continued)

Industry	Pollutants		
	CO_2 (t/million yen)	SOx (t/million yen)	NOx (t/million yen)
25 "Public works administrative organizations" or "public works"	3.81E-01	5.09E-04	7.11E-04
26 Education and research services	3.64E-01	1.24E-03	5.27E-04
27 Medical, health, and social insurance services	3.67E-01	1.26E-03	5.32E-04
28 Other public services	2.79E-01	6.08E-04	2.78E-04
29 Miscellaneous business services	1.26E-01	2.73E-04	1.25E-04
30 Miscellaneous personal services	3.71E-01	8.08E-04	3.69E-04
31 Office supplies	1.36E+00	2.66E-03	5.67E-03
32 Not elsewhere classified	1.36E+00	2.66E-03	5.67E-03
Total emitted by final demand	4.00E-01	2.25E-05	8.38E-04

Table 7.23 Emission coefficients of Japan (2000)

Industry	Pollutants		
	CO_2 (t/million yen)	SOx (t/million yen)	NOx (t/million yen)
1 Agriculture	1.17E+00	7.78E-03	1.41E-02
2 Mining	5.37E-01	8.76E-04	3.63E-03
3 Food products	3.81E-01	1.27E-03	5.05E-04
4 Textiles and clothing	5.01E-01	1.54E-03	1.11E-03
5 Paper and wooden products	1.29E+00	2.86E-03	2.19E-03
6 Chemical products	1.94E+00	2.81E-03	2.71E-03
7 Coal and petroleum products	3.14E+00	4.51E-03	5.15E-03
8 Nonmetallic mineral products	8.14E+00	3.41E-03	1.78E-02
9 Iron and steel	9.58E+00	3.26E-03	4.45E-03
10 Non-ferrous metals	8.94E-01	2.31E-03	2.09E-03
11 Fabricated metal products	3.79E-01	2.07E-04	5.00E-04
12 General machinery	1.35E-01	1.24E-04	2.74E-04
13 Electric machinery	1.10E-01	9.19E-05	1.50E-04
14 Transport equipment, automobiles	1.61E-01	1.34E-04	2.31E-04
15 Precision apparatus	1.35E-01	1.02E-04	1.47E-04
16 Other manufactures	2.46E-01	4.89E-04	4.10E-04
17 Building and construction	1.85E-01	1.45E-04	1.57E-03
18 Electricity, gas	1.96E+01	1.17E-02	1.37E-02
19 Water	4.12E+00	3.12E-03	3.17E-03
20 Wholesale and retail service	1.33E-01	3.19E-04	1.25E-04
21 Finance and insurance	2.85E-02	7.62E-06	3.70E-05
22 Real estate services	4.92E-02	5.11E-05	7.17E-05
23 Transport	4.41E+00	1.94E-02	4.24E-02
24 Communications	7.32E-02	9.20E-05	2.47E-04
25 "Public works administrative organizations" or "public works"	2.97E-01	6.17E-04	7.10E-04
26 Education and research services	3.09E-01	4.08E-04	5.90E-04
27 Medical, health, and social insurance services	2.99E-01	4.58E-04	5.22E-04
28 Other public services	2.82E-01	2.59E-04	3.81E-04
29 Miscellaneous business services	8.21E-02	8.24E-05	1.77E-04
30 Miscellaneous personal services	3.59E-01	2.32E-04	5.27E-04
31 Office supplies	0.00E+00	0.00E+00	0.00E+00
32 Not elsewhere classified	4.10E-01	7.61E-04	1.65E-03
Total emitted by final demand	4.00E-01	2.25E-05	8.38E-04

Table 7.24 Emission coefficients of abatement industries (1997 and 2000)

Industry	Pollutants		
	CO_2 (t/abatement t)	SOx (t/abatement t)	NOx (t/abatement t)
CO_2 abatement	1.04E-02	1.35E-05	1.04E-05
SOx abatement	4.10E-02	5.35E-05	4.12E-05
NOx abatement	2.97E-01	3.87E-04	2.98E-04

Table 7.25 Other parameters (1997) (no.1)

Industry	α	Cg	r	e (million yen)	m (million yen)
1	1.18E-02	0.00E+00	1.68E-03	47, 704	2, 863, 929
2	1.46E-06	0.00E+00	5.46E-04	19, 696	8, 185, 535
3	8.38E-02	0.00E+00	2.47E-03	254, 210	5, 251, 172
4	2.22E-02	0.00E+00	1.44E-03	744, 918	3, 124, 684
5	2.77E-03	0.00E+00	9.30E-03	293, 414	2, 276, 675
6	8.84E-03	0.00E+00	2.29E-03	3, 785, 847	2, 719, 499
7	1.18E-02	0.00E+00	7.66E-04	389, 962	1, 549, 187
8	1.12E-03	0.00E+00	3.89E-04	624, 042	431, 891
9	0.00E+00	0.00E+00	8.64E-04	1, 857, 827	620, 147
10	2.66E-04	0.00E+00	2.54E-04	768, 707	2, 014, 803
11	1.30E-03	0.00E+00	3.45E-03	606, 649	381, 615
12	1.48E-04	0.00E+00	0.1014565	8, 583, 352	1, 738, 693
13	1.99E-02	0.00E+00	0.1182486	14, 701, 490	6, 904, 589
14	2.65E-02	0.00E+00	4.33E-02	10, 910, 630	2, 030, 900
15	2.49E-03	0.00E+00	1.18E-02	1, 649, 784	1, 127, 237
16	1.82E-02	0.00E+00	5.41E-03	1, 713, 718	2, 678, 192
17	0.00E+00	0.00E+00	0.6083234	0	0
18	1.79E-02	0.00E+00	0.00E+00	9103	0
19	4.57E-03	0.00E+00	0.00E+00	1329	0
20	0.12556	0.00E+00	8.22E-02	2, 548, 609	350, 615
21	2.43E-02	0.00E+00	0.00E+00	837, 339	1, 336, 422
22	0.1883766	0.00E+00	0.00E+00	1483	0
23	3.94E-02	0.00E+00	5.92E-03	3, 800, 179	1, 684, 863
24	1.47E-02	0.00E+00	0.00E+00	165, 245	207, 600
25	6.35E-03	0.711813	0.00E+00	0	0
26	2.20E-02	0.2716355	0.00E+00	12, 235	17, 980
27	9.07E-02	1.66E-02	0.00E+00	0	0
28	1.67E-02	0.00E+00	0.00E+00	47, 448	28, 463
29	6.14E-03	0.00E+00	0.00E+00	743, 050	1, 352, 847
30	0.1316646	0.00E+00	0.00E+00	166, 697	676, 155
31	8.72E-04	0.00E+00	0.00E+00	0	0
32	0.00E+00	0.00E+00	0.00E+00	1, 664, 534	1, 697, 230

Table 7.26 Other parameters (1997) (no.2)

τ0	τI	δ	sg	β
0.132047	7.51E-02	0.161112	0.446013	9.99E-02

Table 7.27 Other parameters (2000) (no.1)

Industry	α	Cg	r	e (million yen)	m (million yen)
1	1.17E-02	0.00E+00	1.89E-03	35,409	2,389,381
2	1.16E-06	0.00E+00	1.23E-04	16,195	5,851,915
3	8.05E-02	0.00E+00	3.75E-05	191,225	4,685,017
4	2.12E-02	0.00E+00	1.67E-03	600,681	2,701,730
5	1.01E-02	0.00E+00	9.12E-03	328,976	2,433,897
6	9.81E-03	0.00E+00	−2.27E-03	3,452,786	2,548,435
7	8.76E-03	0.00E+00	−2.60E-04	333,270	1,020,264
8	1.09E-03	0.00E+00	−4.17E-05	525,900	315,491
9	0.00E+00	0.00E+00	−3.89E-04	1,570,634	598,335
10	3.83E-04	0.00E+00	9.86E-05	588,304	1,783,930
11	1.25E-03	0.00E+00	4.12E-03	518,296	310,930
12	1.91E-04	0.00E+00	9.54E-02	7,360,775	1,132,299
13	1.78E-02	0.00E+00	9.99E-02	1.25E+07	4,767,958
14	2.36E-02	0.00E+00	4.79E-02	8,415,179	1,617,064
15	2.06E-03	0.00E+00	1.11E-02	1,267,344	730,471
16	7.87E-03	0.00E+00	5.10E-03	780,606	1,335,364
17	0.00E+00	0.00E+00	0.6468493	0.00E+00	0.00E+00
18	1.59E-02	0.00E+00	0.00E+00	11,384	0.00E+00
19	3.89E-03	0.00E+00	0.00E+00	1662	0.00E+00
20	0.1182744	0.00E+00	7.44E-02	2,103,428	268,915
21	1.98E-02	0.00E+00	0.00E+00	497,588	771,795
22	0.1614781	0.00E+00	0.00E+00	1855	0.00E+00
23	3.66E-02	0.00E+00	5.39E-03	3,393,942	2,033,817
24	1.18E-02	0.00E+00	0.00E+00	47,780	79,906
25	5.63E-03	0.7048947	0.00E+00	0.00E+00	0.00E+00
26	1.96E-02	0.2767956	0.00E+00	9088	15,106
27	7.88E-02	1.83E-02	0.00E+00	0.00E+00	0.00E+00
28	1.52E-02	0.00E+00	0.00E+00	35,244	23,913
29	4.88E-03	0.00E+00	0.00E+00	423,739	786,482
30	0.1181736	0.00E+00	0.00E+00	103,988	537,995
31	7.94E-04	0.00E+00	0.00E+00	0.00E+00	0.00E+00
32	0.00E+00	0.00E+00	0.00E+00	1,338,670	1,377,718

Table 7.28 Other parameters (2000) (no.2)

τ0	τI	δ	sg	β
0.132297	7.20E-02	0.075097	0.466707	1.93E-01

References

Economic Planning Agency of Japan (1999) Input-output table of Japan in 1997 (in Japanese)

Higano Y (1988) Control of environment through quasi-market. Pap RSA 63:43–58

Higano Y (1997) Optimal taxes on the emission of air pollutants in Japan. A paper read at session in honour of professor Hirotada Kohno: environmental policy and sustainability of the 15th Pacific Regional Science Conference, December 8–14th, 1997, Wellington

Society for the Research of the Global Environmental Economics, Experience of Public Bads in Japan: Diseconomies of the Economy which does not Consider the Environment (1991) Go-Do Shuppan, Ltd

Society for the System Research of the Greenhouse Gas Effects on the Global Environment of The Environmental Agency, Government of Japan, The second interim session 1994

Yoshioka K et al (1996) Input-output table for environmental analysis, Institute for Economic and Industrial Studies (Keio Economic Observatory)

Chapter 8
Scenario Input-Output Analysis on the Diffusion of Fuel Cell Vehicles and Alternative Hydrogen Supply Systems Using MRIOT

Mitsuo Yamada, Kiyoshi Fujikawa, and Yoshito Umeda

Abstract According to the 2015 Paris Agreement, Japan is to approach the target of a 26% reduction of its 2013 greenhouse gas (GHG) emissions by 2030. To attain this target, it is necessary to transcend the current fossil energy-based society and shift to a renewable energy-oriented society. Carbon dioxide (CO_2)-free fuels must become the predominant source of energy, in addition to the introduction of energy conservation technologies in each sector of manufacturing, transportation, and business and in households. Fuel cells and hydrogen, therefore, are gaining much attention. Our research group in "Knowledge Hub Aichi" is developing a new hydrogen-generating system, which directly decomposes hydrogen from methane (directly decomposition of methane, DDM) and separates carbon as a solid substance without CO_2 emissions. We have estimated DDM's CO_2 reduction effects and compared them to those in the current steam reforming of methane (SRM) by applying a scenario input-output analysis with a multiregional input-output table (MRIOT). For a certain amount of hydrogen production, DDM directly emits 14.2% of the CO_2 emitted by SRM or 24.5% when considering its indirect effect on the industry. Under the assumption that 800,000 fuel cell vehicles (FCVs) will be diffused in Japan before 2030, the total reduction of CO_2 from DDM is estimated as 21.8% more than that from SRM, when FCVs replace conventional vehicles. The vehicle substitution requires a regional concentration of vehicle production

M. Yamada (✉)
School of Economics, Chukyo University, Nagoya, Japan
e-mail: yamada@mecl.chukyo-u.ac.jp

K. Fujikawa
Applied Social System Institute of Asia, Nagoya University, Nagoya, Japan

Institute of Economics, Chukyo University, Nagoya, Japan
e-mail: fujikawa@gsid.nagoya-u.ac.jp

Y. Umeda
Toho Cryogenics Co., Ltd., Nagoya, Japan
e-mail: umeda@tohogas.co.jp

K. Nakayama, Y. Miyata (eds.), *Theoretical and Empirical Analysis in Environmental Economics*, New Frontiers in Regional Science: Asian Perspectives 34, https://doi.org/10.1007/978-981-13-2363-8_8

in the Aichi Prefecture, but then the production of Aichi would increase with the resulting additional CO_2 emission. DDM's introduction suppresses the increase of CO_2 emissions in the industry.

Keywords Scenario input-output analysis · Fuel cell vehicles · Hydrogen production technology · Carbon dioxide emissions · Direct decomposition of methane method · Steam reforming of methane method · Multiregional input-output table

8.1 Introduction

The Paris Agreement,[1] the new framework for global environmental measures from 2020 onward, concluded the United Nations Framework Convention on Climate Change (UNFCC) COP21 in 2015 and came into effect in 2016. According to the agreement, Japan aims to increase its target of reducing greenhouse gas emissions to 26% below 2013 levels by 2030. On the other hand, Japan's greenhouse emissions have increased greatly through the influence of the nuclear plant closure after the East Japan great earthquake. To simultaneously attain the Paris target and ensure energy security, it is essential for Japan to transcend the current fossil energy-based (petroleum, coal, and natural gas) society and shift to a renewable energy-oriented society as most of Japan's primary energy depends on overseas fossil fuels. Carbon dioxide (CO_2) emission-free fuels must become the dominant form, in addition to introducing energy conservation technologies in each sector of manufacturing, transportation, business, and households. Thus, as is mentioned later, fuel cells and hydrogen[2] are gaining much attention.

The engineering team of our research group is developing a new hydrogen-generating system that directly decomposes hydrogen from methane and separates carbon as a solid substance, with zero CO_2 emissions. Our task is to evaluate the economic and environmental effects of CO_2 reduction on the entire economy when the new system is introduced in society. Here, we conduct a simulation analysis to estimate the CO_2 reduction from our new hydrogen production system and compare it to that from the current steam reforming method by applying a scenario input-output analysis. In implementing this simulation analysis, we assume a certain volume of diffusion of fuel cell vehicles in the future and use it as a reference case for the simulation analysis.

---

[1] The feature of this agreement is that each country sets an individual voluntary reduction target on greenhouse gas (GHG) emissions called Nationally Determined Contributions (NDCs).

[2] Hydrogen, a source of energy for a fuel cell, can be generated in various ways. Representative examples include extraction from fossil fuels or electrolysis of water. Currently, it is already in practical use to generate hydrogen by steam reforming from petroleum, natural gas, or gasification of coal. These methods, however, have the disadvantage that they emit CO_2 in the process of generating hydrogen.

Many extended input-output models have been applied in the energy and environmental fields. According to Miller and Blair (2009), since the late 1960s, many researchers theorized that the input-output framework could be extended to account for environmental pollution generation and abatement processes associated with inter-industry activity. Leontief (1970) provided a key methodological extension that has since been applied widely and extended further. In relation to CO_2 emissions, there is an accumulation of studies on carbon footprint, as in Nansai et al. (2009), Wiebe et al. (2012), and Usubiaga and Acosta-Fernandez (2015).

Cantono et al. (2008) discussed the effects of public transportation using hydrogen and fuel cell buses through environmental input-output analysis. When hydrogen was produced by steam reforming of methane, CO_2 emissions were not completely reduced, even though fuel cell buses did not emit CO_2 during operation. They suggested the use of hydrogen in fuel cell buses, as it is environmentally desirable, especially if it is accompanied by the employment of renewable sources, CO_2 capture, or both.

Keipi et al. (2018) compared the costs of hydrogen production using thermal decomposition of methane to those of steam methane reforming and water electrolysis in the current and potential future market environments. They estimated costs from engineering-based information and not from input-output tables. They found that thermal decomposition of methane would be most suited for on-site demand-driven hydrogen production in small- and medium-scale operations and would also be economically competitive with steam reforming. Thermal decomposition of methane has the advantage of feedstock availability via the current natural gas infrastructure, whereas electrolysis is highly dependent on the cost and availability of renewable electricity.

Our study is an environmental input-output analysis, focusing on the economic effects of new hydrogen production technologies. As reference, we focus on hydrogen demand from fuel cell vehicles—passenger cars bought by consumers—and not fuel cell buses for public transportation. Further, we compare the new hydrogen production technology from methane decomposition to steam reforming, which is currently in operation. Electrolysis is omitted from the scope of this study, because the cost information, which depends on the power sources, is not obtainable.

In our input-output model, there are multiple technologies in hydrogen production for one commodity, and so the row size differs from the column size of the input coefficient matrix, which requires some arrangement to obtain the Leontief inverse matrix. Here, we introduce a weighted average of the plural technologies. The weights, which are given exogenously as a scenario, indicate the state of technology choice. Similar analyses appear in Yoshioka and Suga (1997), Wang (2016), and Fujikawa and Wang (2017).

This paper is organized as follows: In the next section, we describe the Hydrogen Basic Strategy in Japan and the current situation of next-generation vehicles. Section 8.3 outlines the input-output scenario analysis model, Section 8.4 describes data and assumptions, and Sect. 8.5 discusses the analytical results of the scenario input-output analysis. We summarize these results in Sect. 8.6.

8.2 Hydrogen Basic Strategy and the Current Situation of Fuel Cell Vehicles (FCVs) in Japan

8.2.1 Hydrogen Basic Strategy in Japan

For the first time, in April 2014, hydrogen energy was recognized as one of the promising substitutes for fossil energies in the Fourth Energy Basic Plan, where a hydrogen society is considered a viable future option for Japan. Further, after attending the Hydrogen and Fuel Cell Strategy Conference organized by personnel in the industry, universities and the government issued the "Hydrogen and Fuel Cell Strategy Roadmap" in June of the same year. This Roadmap presented the following three phase programs to realize a hydrogen society, paying attention to the required period to overcome the technical and economic problems.

Phase 1: Rapid Expansion of Hydrogen Use (Present)

Fixed fuel cell batteries and fuel cell vehicles should be diffused further in order for Japan to acquire a global market share in the field of hydrogen and fuel cells.

Phase 2: Introduction of Full-Fledged Hydrogen Power Generation and Establishment of a Large-Scale Hydrogen Supply System (Before the Latter Half of the 2020s)

The secondary energy supply/demand structure should be revamped to one that includes hydrogen in addition to electricity and heat through further expansion of hydrogen demands and utilization of unused hydrogen resources.

Phase 3: Establishment of CO_2-Free Hydrogen Supply System (Before Approximately 2040)

A CO_2-free hydrogen supply system should be established by a combination of CCS and hydrogen production technology with renewable electricity.

The Hydrogen and Fuel Cell Strategy Roadmap was revised in March 2016, based on the progress of mainstream social recognition on hydrogen usage, where quantity targets were introduced for household fuel cell (*Enefarm*) or FCVs/hydrogen stations.

The first "Ministerial Conference on Renewable Energy and Hydrogen" was held in April 2017, for a discussion on the expansion of renewable energy and realization of a hydrogen-based society. The conference decided to issue a Hydrogen Basic Strategy in 1 year, in order for Japan to be an early adopter of a hydrogen-based society, and this policy was described in "Investment for the Future Strategy 2017" (with cabinet approval on June 9, 2017). The Hydrogen Basic Strategy was adopted in a cabinet meeting organized by the ministers in charge of renewable and hydrogen energy held in December 26, 2017, where "3E + S"—Energy security, Economic efficiency, Environment, and Safety—was set as the main perspective of the Japanese energy policy. The government recognizes that the risk of global warming, especially in the latter half of this century, is so large that the damage to future generations is expected to be very severe. The government also knows that it is not enough to continue the current energy-saving efforts to reduce CO_2 emissions or to stabilize the CO_2 density in the air. The strategy emphasizes that a positive introduction of innovative technology is indispensable, and general mobilization of all the measures is required in order to realize a "3E + S" society. Then, hydrogen gained attention as a new, promising energy carrier for the future.

The Hydrogen Basic Strategy proposes three uses of hydrogen: as an electricity power source, energy for industrial processes, and power for transportation. Here, let us introduce how the government considers hydrogen use in transportation.

(a) *FCVs and Hydrogen Stations*

Diffusion of FCV and hydrogen stations is the kernel of hydrogen use in transportation. The target for the diffusion of FCVs is 40,000 units by 2020, 200,000 units by 2025, and 800,000 units by 2030, whereas the target for hydrogen stations is 160 stations by 2020 and 320 stations by 2025. Moreover, the hydrogen station business should become profitable by the latter half of the 2020s. A price reduction of low-end or volume zone FCVs by extension of mass production, as well as a decrease in the hydrogen supply cost (cost competitiveness with gasoline), is required to accomplish the aforementioned goal. It is indispensable to reform regulations on hydrogen supply, to enhance technological development of hydrogen energy, and to install and utilize hydrogen stations strategically with the cooperation of the public and private sectors in order to foster economically autonomic FCV makers and the hydrogen station industry.

(b) *Renewable Energy-Origin Hydrogen Station*

The renewable energy-origin hydrogen station has such advantages as an overall greenhouse gas (GHG) reduction in the hydrogen manufacturing process, advancement of local production and local consumption of renewable energies, stimulation of local hydrogen demand, and space-saving for hydrogen stations. Furthermore, a renewable energy origin hydrogen station has an important meaning from the viewpoint of improving the societal acknowledgment of renewable energy-origin hydrogen or the societal acceptability of renewable energy-origin hydrogen, especially in the early days of diffusion of FCVs. The hydrogen filling pressure in a renewable energy-origin hydrogen station is currently 35 MPa, which is not enough for FCVs. A 70 MPa renewable energy-origin hydrogen station, however, will operate for FCVs in 2018.

(c) *Fuel Cell Bus (FCB)*

Diffusion of FCBs as one of the public transportation facilities is meaningful largely from the viewpoint that it promotes public understanding, since it provides opportunities for many people to be able to become familiar with hydrogen and to experience a hydrogen-powered society. Because a short charging time, long cruising range, and flexibility of the route are very important for public buses, FCB is superior to a battery electric vehicle (BEV). In addition, FCB is expected to be an emergency power source in the case of disasters since it has a large external power feeding function for 4–5 days in one refuge. FCB commercial operation on a regular route has already started in 2017, and the target of the diffusion of FCBs is 100 units by 2020 and 1200 units by 2030.

(d) *Fuel Cell Forklift (FCF)*

The fuel cell forklift (FCF) has an advantage in terms of filling time and amount of CO_2 emissions compared to those of a battery electric forklift (BEF). On the other hand, the high initial cost and fuel cell availability is a current challenge for FCFs compared to BEFs. Confined to large-scale forklift users, 120,000 forklifts are currently in operation. They are a source of a large hydrogen demand, approximately equivalent to that of 360,000 units of low-end FCVs. The distribution of FCFs has already started in 2016, and the target of their diffusion is 500 units by 2020 and 10,000 units by 2030.

(e) *Fuel Cell Tracks (FCT) Development*

The CO_2 emissions from trains are large since they account for 36% of those of the entire transportation section. Therefore, the room for achieving low carbonization in powering train tracks is also large. What is required to develop zero-emission FCT is extending the cruising range and decreasing the weight of the power train (devices for drive). FCT are superior to electric tracks in the area of 100 km or more, considering the weight of the hydrogen tank and battery in one unit. The potential demand of hydrogen for tracks is expected to be large since the number of industrial trains is 3.2 million or more, far larger than the number of buses (230,000).

(f) *Introduction of a Fuel Cell Ship (FCS)*

It will be necessary to reduce CO_2 emissions by advancing electrification, including the use of the fuel cell in the future, although low carbonization is difficult in ships in the transportation field. Therefore, the introduction of the fuel cell can be launched for small ships such as pleasure boats, passenger/sightseeing boats, and fishing boats since the silent operation of the fuel cell is a great advantage for them. Security guidelines for fuel cell ships should be formulated as soon as possible.

8.2.2 The Current Situation of FCVs in Japan

Table 8.1 shows the trends in Japan's recent greenhouse gas (GHG) and sectoral CO_2 emissions. The amount of greenhouse gas (GHG) emissions was 1322 million tons in FY2016, down by 4.6% and 6.2% compared to those in FY2005 and FY2013, respectively. CO_2 emissions account for 1222 million tons, or 92.4% of the total GHG emission. Of the CO_2 emissions, 93.4%, or 1144 million tons, originated from energy sources, whereas the remaining 78 million tons are of non-energy origins. The industrial sector (energy origin) accounts for 34.2% of the total CO_2

Table 8.1 GHG and CO_2 emissions by sector (million t-CO_2 equivalent)

	Fiscal year	1990	2005	2013	2015	2016 (preliminary)	Growth rate from 2005	Growth rate from 2013	Growth rate from 2015
GHG	Total	1,277	1,386	1,409	1,325	1,322	−4.6%	−6.2%	−0.2%
	Carbon dioxide (CO_2)	1,166	1,297	1,316	1,228	1,222	−5.8%	−7.1%	−0.5%
	Methane (CH_4) and others	111	89	93	97	100	12.4%	7.5%	3.1%
Carbon dioxide, total		1,166	1,297	1,316	1,228	1,222	−5.8%	−7.1%	−0.5%
Energy origin	Subtotal	1,070	1,206	1,235	1,150	1,144	−5.1%	−7.4%	−0.5%
	Industry sector	502	468	463	435	418	−10.7%	−9.7%	−3.9%
	Transportation sector	207	245	224	217	215	−12.2%	−4.0%	−0.9%
	Business and other sectors	130	217	244	231	219	0.9%	−10.2%	−5.2%
	Household sector	131	175	205	184	179	2.3%	−12.7%	−2.7%
	Energy conversion sector	99.8	100	100	82	113	13.0%	13.0%	37.8%
Non-energy origin	Subtotal	95.6	91.8	80.9	78.3	78	−15.0%	−3.6%	−0.4%
	Production processes and use of products	65.1	55.6	48	46.1	45.7	−17.8%	−4.8%	−0.9%
	Waste	24	31.7	29.4	28.8	29	−8.5%	−1.4%	0.7%
	Others	6.5	4.5	3.5	3.3	3.3	−26.7%	−5.7%	0.0%
GHG	Total	100.0%	100.0%	100.0%	100.0%	100.0%			
	Carbon dioxide (CO_2)	91.3%	93.6%	93.4%	92.7%	92.4%			
	Methane (CH_4) and others	8.7%	6.4%	6.6%	7.3%	7.6%			
Carbon dioxide, total		100.0%	100.0%	100.0%	100.0%	100.0%			
Energy origin	Subtotal	91.8%	93.0%	93.8%	93.6%	93.6%			
	Industry sector	43.1%	36.1%	35.2%	35.4%	34.2%			
	Transportation sector	17.8%	18.9%	17.0%	17.7%	17.6%			
	Business and other sectors	11.1%	16.7%	18.5%	18.8%	17.9%			
	Household sector	11.2%	13.5%	15.6%	15.0%	14.6%			
	Energy conversion sector	8.6%	7.7%	7.6%	6.7%	9.2%			
Non-energy origin	Subtotal	8.2%	7.1%	6.1%	6.4%	6.4%			
	Production processes and use of products	5.6%	4.3%	3.6%	3.8%	3.7%			
	Waste	2.1%	2.4%	2.2%	2.3%	2.4%			
	Others	0.6%	0.3%	0.3%	0.3%	0.3%			

Source: "Japan's greenhouse gas emissions" (Ministry of Environment, Japan, 2017)

emissions whereas the business and other sectors for 17.9%, the transport sector 17.6%, the household sector 14.6%, and the energy conversion sector 9.2%.

In the transportation sector, next-generation vehicles have to play an important role in reducing CO_2 emissions. Next-generation vehicles include electric vehicles (EVs), plug-in hybrid vehicles (PHVs), FCVs, and hybrid vehicles (HVs). Both the number of HVs owned and their sales volume are overwhelmingly large. The number of HVs sold was 1.337 million units in FY2016 and that of vehicles owned at the end of FY2016 stood at 6.971 million units. The total number of EVs and PHVs sold was 13,000 units, that of EVs owned was 70,000 units, and that of PHVs 89,000 units. FCV sales started in 2014 and 1807 units were sold by the end of FY2016.[3]

Figure 8.1 represents well-to-wheels-based CO_2 emissions for each vehicle. For gasoline vehicles, these are 147 g-CO_2/km, for diesel vehicles slightly lower at 132 g-CO_2/km, and for HVs 95 g-CO_2/km. The amount of gasoline refueling for PHVs is almost equal to that for an HV, and the emissions drop to 102 g-CO_2/km and 55 g-CO_2/km, respectively, when they are charged. In EVs, emissions depend on the mix of power sources; they were 55 g-CO_2/km in 2009 and 77 g-CO_2/km in 2012 when nuclear power plants were shut down due to the Great East Japan Earthquake of 2011. In contrast, when electricity generated from photovoltaic power is used, almost no CO_2 is generated; the emissions are 1 g-CO_2/km.

Further, FCVs depend on the hydrogen production technology. Their emissions are 79 g-CO_2/km when hydrogen is used by on-site reforming of gas and 78 g-CO_2/km for off-site reforming of natural gas. These amounts are not very different from those of EVs, depending on the current mix of power sources. Gas-reforming technologies are currently established to produce hydrogen. For on-site alkaline water electrolysis with solar power, emissions become considerably low at 14 g-CO_2/km.

Below, we evaluate the effect of another hydrogen production technology called methane direct decomposition method, which produces hydrogen from methane with far lower CO_2 emissions and carbon as solid coal.

8.3 Methods: Scenario Input-Output Analysis

When there is more than one activity (production technology) for a product in the input-output analysis, there appears to be difficulty in handling technology selection among the several activities. One approach is solving the equation with additional constraints on the input coefficient matrix.[4] The electric power generation sector is a typical example but involves only one product. In Japan's input-output table, there

[3]These data are obtained from Next Generation Vehicle Promotion Center (2018).

[4]Input-output analysis with a single product produced by multiple activities appears in Yoshioka and Suga (1997), Wang (2016), and Fujikawa and Wang (2017).

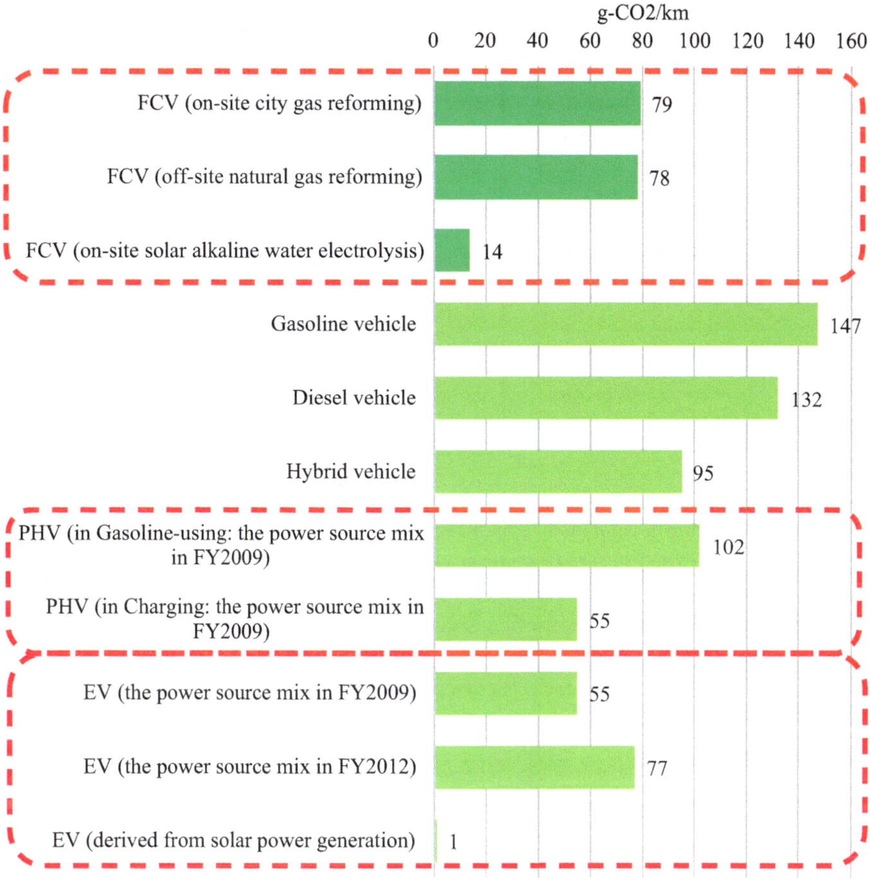

Fig. 8.1 Comparison of CO_2 emissions (well-to-wheels JC08 mode). Source: "Report on overall efficiency and GHG emissions by type of vehicle" (Japan Automobile Research Institute, March, 2011)

are three activities: (a) nuclear power, (b) fire power, and (c) hydraulic power and other activities. Figure 8.2 shows the input-output table.

If we change the composition of these activities, the environmental load and economic effects also change. In this study, the technology for hydrogen production consists of two methods: the conventional one and direct decomposition of methane. If the input and energy utilization structures differ for each hydrogen production technology, the environmental load and economic effects will also change if altering the composition.

This input-output model is expressed as follows:

$$\begin{aligned} \mathbf{A}_{11}\mathbf{x}_1 + \mathbf{A}_{12}\mathbf{z} + \mathbf{f}_1 = \mathbf{x}_1 \\ \mathbf{A}_{21}\mathbf{x}_1 + \mathbf{A}_{22}\mathbf{z} + \mathbf{f}_2 = \mathbf{x}_2 \end{aligned} \tag{8.1}$$

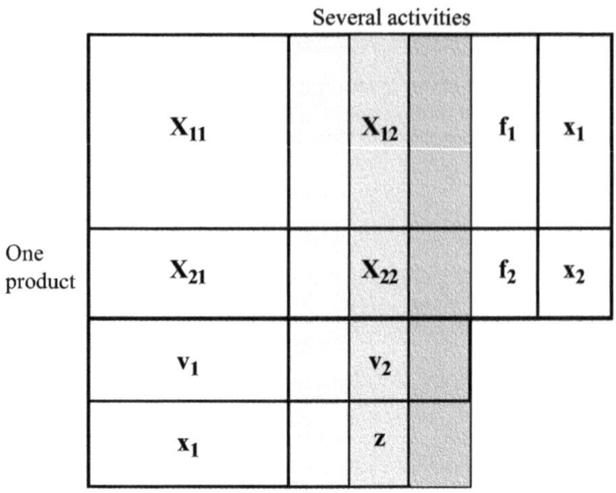

Fig. 8.2 Input-output table with non-square input transaction matrix. Source: Illustrated by Authors

where $\mathbf{x_1}$ is the output of sector 1, excluding the power sector, $\mathbf{x_2}$is the output of the power sector, and $\mathbf{z}$ is the power generated by different technologies. $\mathbf{A_{ij}}$ is the input coefficient matrix of sector j from sector i, and $\mathbf{f_i}$ is the final demand for sector i.

Since there are multiple activities, the input coefficient matrix does not become a square matrix, and it is impossible to obtain the usual Leontief inverse matrix. Therefore, the following scenario (restriction) is added.

$$\begin{bmatrix} z_1 \\ z_2 \\ z_3 \end{bmatrix} = \begin{bmatrix} \alpha_1 \\ \alpha_2 \\ 1 - \alpha_1 - \alpha_2 \end{bmatrix} \mathbf{x_2}$$

$$\mathbf{z} = \mathbf{c}\ \mathbf{x_2} \tag{8.2}$$

Here, vector $\mathbf{c}$ on the right-hand side represents the scenario. These proportions will be given exogenously. Typically, if vector $\mathbf{c}$ changes, the required production volume also changes. Substituting Eq. (8.2) into Eq. (8.1), we obtain

$$\begin{bmatrix} \mathbf{A_{11}} & \mathbf{A_{12}c} \\ \mathbf{A_{21}} & \mathbf{A_{22}c} \end{bmatrix} \begin{bmatrix} \mathbf{x_1} \\ \mathbf{x_2} \end{bmatrix} + \begin{bmatrix} \mathbf{f_1} \\ \mathbf{f_2} \end{bmatrix} = \begin{bmatrix} \mathbf{x_1} \\ \mathbf{x_2} \end{bmatrix} \tag{8.3}$$

The input coefficient matrix becomes square, and the Leontief inverse matrix can be obtained as follows:

$$\begin{bmatrix} \mathbf{x_1} \\ \mathbf{x_2} \end{bmatrix} = \begin{bmatrix} \mathbf{I} - \mathbf{A_{11}} & -\mathbf{A_{12}c} \\ -\mathbf{A_{21}} & \mathbf{I} - \mathbf{A_{22}c} \end{bmatrix}^{-1} \begin{bmatrix} \mathbf{f_1} \\ \mathbf{f_2} \end{bmatrix} \tag{8.4}$$

In the power sector, vector **c** represents the composition of the power supply in terms of nuclear power, thermal power, hydroelectric power, and others. It is possible to estimate the influence obtained by changing the composition of the power supply.

In the case of hydrogen production, by changing the proportion of the conventional type and of the direct decomposition of methane method, it is possible to see how much they influence CO_2 emissions. In addition, it is possible to simulate how much CO_2 emissions differ by the choice of the production technologies at the common final demand level.

If we extend this model to two regions ("a" and "r"), the multiregional input-output model corresponding to Eq. (8.3) is expressed as follows:

$$\begin{bmatrix} \mathbf{A}_{11}^{aa} & \mathbf{A}_{12}^{aa}\mathbf{c}^{a} & \mathbf{A}_{11}^{ar} & \mathbf{A}_{12}^{ar}\mathbf{c}^{r} \\ \mathbf{A}_{21}^{aa} & \mathbf{A}_{22}^{aa}\mathbf{c}^{a} & \mathbf{A}_{21}^{ar} & \mathbf{A}_{22}^{ar}\mathbf{c}^{r} \\ \mathbf{A}_{11}^{ra} & \mathbf{A}_{12}^{ra}\mathbf{c}^{a} & \mathbf{A}_{11}^{rr} & \mathbf{A}_{12}^{rr}\mathbf{c}^{r} \\ \mathbf{A}_{21}^{ra} & \mathbf{A}_{22}^{ra}\mathbf{c}^{a} & \mathbf{A}_{21}^{rr} & \mathbf{A}_{22}^{rr}\mathbf{c}^{r} \end{bmatrix} \begin{bmatrix} \mathbf{x}_1^{a} \\ \mathbf{x}_2^{a} \\ \mathbf{x}_1^{r} \\ \mathbf{x}_2^{r} \end{bmatrix} + \begin{bmatrix} \mathbf{f}_1^{aa} \\ \mathbf{f}_2^{aa} \\ \mathbf{f}_1^{ra} \\ \mathbf{f}_2^{ra} \end{bmatrix} + \begin{bmatrix} \mathbf{f}_1^{ar} \\ \mathbf{f}_2^{ar} \\ \mathbf{f}_1^{rr} \\ \mathbf{f}_2^{rr} \end{bmatrix} = \begin{bmatrix} \mathbf{x}_1^{a} \\ \mathbf{x}_2^{a} \\ \mathbf{x}_1^{r} \\ \mathbf{x}_2^{r} \end{bmatrix} \tag{8.5}$$

where $\mathbf{x_{is}}$ is the output of sector i in region s, $\mathbf{A_{ijst}}$ the input coefficient matrix of sector j in region t from sector i in region s, and $\mathbf{f_{ist}}$ the final demand for sector i in region t from region s. $\mathbf{c}^{s}$ is the proportions of products supplied by the different technologies in region s. Then, the solution for Eq. (8.5) is as follows:

$$\begin{bmatrix} \mathbf{x}_1^{a} \\ \mathbf{x}_2^{a} \\ \mathbf{x}_1^{r} \\ \mathbf{x}_2^{r} \end{bmatrix} = \begin{bmatrix} \mathbf{I}-\mathbf{A}_{11}^{aa} & -\mathbf{A}_{12}^{aa}\mathbf{c}^{a} & -\mathbf{A}_{11}^{ar} & -\mathbf{A}_{12}^{ar}\mathbf{c}^{r} \\ -\mathbf{A}_{21}^{aa} & \mathbf{I}-\mathbf{A}_{22}^{aa}\mathbf{c}^{a} & -\mathbf{A}_{21}^{ar} & -\mathbf{A}_{22}^{ar}\mathbf{c}^{r} \\ -\mathbf{A}_{11}^{ra} & -\mathbf{A}_{12}^{ra}\mathbf{c}^{a} & \mathbf{I}-\mathbf{A}_{11}^{rr} & -\mathbf{A}_{12}^{rr}\mathbf{c}^{r} \\ -\mathbf{A}_{21}^{ra} & -\mathbf{A}_{22}^{ra}\mathbf{c}^{a} & -\mathbf{A}_{21}^{rr} & \mathbf{I}-\mathbf{A}_{22}^{rr}\mathbf{c}^{r} \end{bmatrix}^{-1} \left[\begin{bmatrix} \mathbf{f}_1^{aa} \\ \mathbf{f}_2^{aa} \\ \mathbf{f}_1^{ra} \\ \mathbf{f}_2^{ra} \end{bmatrix} + \begin{bmatrix} \mathbf{f}_1^{ar} \\ \mathbf{f}_2^{ar} \\ \mathbf{f}_1^{rr} \\ \mathbf{f}_2^{rr} \end{bmatrix} \right] \tag{8.6}$$

8.4 Data and Assumptions

In our analysis below, we used a multiregional input-output table, which has 2 regions—the Aichi Prefecture and the rest of Japan (ROJ)—and 188 sectors. In Aichi Prefecture, the "Mirai" FCV is currently produced. We compiled this table based on the 2011 national input-output table (benchmark table) and the employment table and the 2011 Aichi Prefecture's input-output table and its employment table. The information for ROJ was obtained by subtracting the values of the Aichi Prefecture from those of Japan. As for CO_2 emissions, we obtained the sectoral CO_2 emissions given by the National Institute for Environmental Studies' 3EID (2018) corresponding to the 2011 Input-Output Table.

We examine the impact of the following three aspects:

1. By introducing FCVs instead of conventional vehicles, there appears to be a fuel substitution effect from gasoline—a conventional fuel (assuming gasoline vehicles)—to hydrogen, the fuel of FCVs.
2. Conventional vehicles and FCVs have different economic effects on the automobile production process because the input structure for manufacturing each type of automobile is different.
3. By taking two hydrogen production technologies, namely, the steam reforming of methane method (SRM) and the direct decomposition of methane method (DDM), different economic and environmental effects are expected because these hydrogen production technologies have different input structures.

For the hydrogen production sector, there will be different activities for one product, and it is necessary to generate the composition proportions externally as a scenario.

8.4.1 Input Structure of the Automobile Sector

According to the Hydrogen and Fuel Cell Strategic Roadmap, we assume here that the spread of FCVs will proceed at a rate of 53,333 vehicles per year (for a total of 800,000 vehicles in 15 years).[5] Additionally, we assume that the prices of major FCV components decline mainly because of the scale effect of the production, and the purchase price of an FCV decreases to 369.6 million yen, 51.1% of the current price, or 723.6 million yen in the case of "Mirai." Then, we estimate that FCV annual sales revenue is 197.14 billion yen. Since it is assumed that the price of conventional vehicles replaced by FCVs is the same, the sales value of conventional vehicles is 197.14 billion yen.

We estimate the input structure of FCV as follows: We refer to the input of a gasoline vehicle. First, based on an ordinary-sized car's input structure, we break down the total cost into major input expenses and indirect expenses. Raw material and parts costs to produce FCVs are modified, based on the costs of parts necessary for FCV production.[6] Inputs for production machinery, electric machinery, and electronic components of FCVs have increased, and, conversely, input for transportation machines (automobile parts) is decreasing. We finally estimate sectoral inputs by adding two estimated types of costs, namely, major and indirect costs.

[5]According to the report, the Hydrogen and Fuel Cell Strategic Roadmap, released in March 2016 by the Hydrogen and Fuel Cell Strategy Council of METI, Japan, the number of fuel cell vehicles (stock base) is projected to be 800,000 units by 2030. The price of FCVs to be realized would be equivalent to that of a hybrid vehicle price by 2025. In addition, the plan is to set up 900 hydrogen stations by 2030. The price of hydrogen is equal to or less than cost of fuel for hybrid vehicles.

[6]Necessary data are obtained from Chubu Region Institute for Social and Economic Research (2015).

Then, we estimate total fuel purchases for gasoline vehicles and for FCVs in the market. Here, if the average mileage[7] of a vehicle is 8000 km/year and fuel efficiency[8] is 10 km/l, then its annual gasoline consumption will be 800.0 l/year. If gasoline price[9] is 137.8 yen/l, then the total value of annual gasoline consumption is estimated at 58.78 billion yen. On the other hand, since the tank capacity of an FCV is 5 kg of hydrogen and its cruising distance 650 km, hydrogen fuel efficiency is calculated to be 130 km/kg. Assuming the same average mileage of 8000 km/year, hydrogen consumption is 61.54 kg/year. Therefore, if the price of hydrogen is 1080 yen/kg, the total value of hydrogen consumption is estimated at 35.45 billion yen.

Total gasoline consumption for 800,000 units is 640,000 kl, and consumption value is 881.73 billion yen. Hydrogen consumption is 49,230,769 kg, and consumption value is 531.69 billion yen. Thus, CO_2 emissions due to consumption of gasoline are 1,486,080 t-CO_2,[10] and the CO_2 emission coefficient is estimated as 16.854 t-CO_2/million yen.

8.4.2 *Hydrogen Production and Input Structure*

We compare the two technologies of hydrogen production from methane: SRM and DDM. For SRM, the chemical reaction that represents producing hydrogen from methane, the primary component of Japanese residential and industrial natural gas, is

$$CH_4 + 2H_2O \rightarrow 4H_2 + CO_2 \tag{8.7}$$

The hydrogen production process generates CO_2. If hydrogen is produced without generating CO_2 from the methane gas, CO_2 emissions from FCVs can be significantly reduced. This is possible through another hydrogen production technology, namely, DDM. Similarly, when using methane, the relevant chemical reaction is

$$CH_4 \rightarrow 2H_2 + C \tag{8.8}$$

In this case, instead of CO_2, solid carbon (C) is generated. Once this technology is established, there is a possibility that CO_2 emissions from FCVs will approach those from on-site alkaline water electrolysis with solar power.

[7]The average mileage of conventional vehicles is obtained from the survey data in the Next-Generation Vehicle Promotion Center report on the diffusion of clean-energy vehicles in 2017.

[8]Fuel efficiency of gasoline vehicles in 2015 is calculated from the Fuel Consumption Statistics of Japan's Ministry of Land, Infrastructure, and Transport.

[9]This value is from the Ministry of Resources and Energy, 2015.

[10]The CO_2 emission coefficient for gasoline is 2.322 kg-CO_2/l.

In the case of SRM, according to Eq. (8.7), with 250 mol of material methane and 46.213 mol of heating methane, a total of 296.213 mol are required for producing 1000 mol of hydrogen, with 296.213 mol of CO_2 emissions. On the other hand, with DDM, according to Eq. (8.8), 500 mol of material methane and 41.951 mol of heating methane, that is, a total of 541.951 mol, are required for producing 1000 mol of hydrogen.[11] In this process, the CO_2 generated is only 41.951 mol due to the combustion of the heating methane, and 500 mol of (solid) carbon are generated. For producing the same amount of hydrogen, DDM requires 1.83 times more methane (as a raw material and for heating) than SSR but generates only 14.2% of the CO_2 than SSR does, as stated below.

Methane for materials costs can be calculated from the engineering relationship and methane price. Capital depreciation was calculated assuming there are 900 hydrogen stations, at a construction cost of 500 million yen per site, with a service life of 20 years. Indirect expenses were obtained as the annual expenses of 20 million yen per station, an estimate of the Ministry of Economy, Trade and Industry (METI), Japan. For both technologies, a transportation margin for hydrogen is required in case of off-site production, but not for on-site production. The coefficient of CO_2 emissions is 0.855 t-CO_2/million yen for DDM and 6.034 t-CO_2/million yen for SRM. With DDM, CO_2 emissions are as low as 14.2% of those with SRM.

The input structure of hydrogen production was estimated as follows: First, the total cost was divided into direct expenses, such as methane for materials, and other indirect expenses. The former are estimated from the engineering relationship of the main materials and energy, and the latter are obtained by referring to the input structure of the headquarters sector as in the case of FCV. Finally, we summed up both and set them as sectoral inputs for hydrogen production. The main difference in the input structure of the two technologies is the amount of methane used as raw material and as a source of heat. DDM needs roughly double the methane input of SRM. However, the former has a larger advantage, since its CO_2 emissions are far lower. The presence or absence of a transport margin also depends on whether production is on-site or off-site.

8.5 Simulation Results

A simulation is conducted for the following cases:

1. Gasoline vehicle purchases: Purchase of 53,333 gasoline vehicles per year, at the same price as FCVs. Those vehicles are supplied regionally, proportional to each region's actual production.
2. FCV purchases: Purchase of 53,333 FCVs per year. The FCVs are supplied only from the Aichi Prefecture, where the factory is located.

---

[11] SRM and DDM require 41.2 kJ/mol-H_2 and 37.4 kJ/mol-H_2, respectively. Methane's heat value is 39.8 MJ/Nm^3, and its molar volume is 22.4 l; then, the volume of heating methane is calculated.

3. Gasoline purchase: Annual gasoline purchase for the gasoline vehicles in case 1. Assuming that the gasoline vehicles purchased are distributed regionally proportionally to the vehicles' demand, gasoline purchase is also distributed in the same way.
4. Hydrogen purchase: Annual purchase of fuel hydrogen for FCVs in case 2. The regional distribution of hydrogen purchase is done in the same way as in case 3.[12]
5. Replace gasoline vehicles by FCVs (subtract case 2 from case 1).
6. Replace gasoline by hydrogen (subtract case 4 from case 3).

Table 8.2 shows the final demand, induced production amount, gross value added, numbers of workers, and CO_2 emissions for cases 1 through 6, when hydrogen is produced only by SRM. Furthermore, the CO_2 emissions from industries (endogenous sectors) and household sectors are included. Table 8.2 also shows not only national values but also regional ones, for Aichi Prefecture and the ROJ.

Although the final demand for gasoline vehicles and FCVs is the same, the economic ripple effect of FCVs becomes relatively small compared to that of gasoline vehicles. Production of gasoline vehicles uses more automobile parts, the sector that has a greater ripple effect. On the other hand, FCVs use more electric machinery parts, electronic parts, and industrial machinery. For this reason, the effect of substitution of FCVs for gasoline vehicles in case 5 is negative for production and value added, although the effect on employment is almost zero. The substitution also has the effect of increasing CO_2 emissions by 18,865 t-CO_2.

Under the assumption of same mileage per year, gasoline consumption is higher than that of hydrogen. Therefore, replacing energy from gasoline by energy hydrogen reduces final demand, resulting in a negative impact on production and employment, although the effect on gross value added is slightly positive. The CO_2 emissions in the industrial sector increase by 7891 t-CO_2, because the hydrogen produced by SRM brings much CO_2 emissions. However, in the household sector, consumption of gasoline directly results in 99,072 t-CO_2 emissions, but by replacing it with hydrogen, the same amount of CO_2 reduction is achieved. Overall, these amounts decrease by 108,335 t-CO_2.

In cases 1 through 6 of Table 8.2, we evaluate the effect of replacing gasoline vehicles by FCVs on production and the changing effect of fuel purchases required for using vehicles per year. However, vehicles, as durable consumer goods, can be used for a certain period of time, during which fuel purchase is required. According to the Ministry of Land, Infrastructure, Transport and Tourism statistics, a passenger car's average duration of usage in 2017 was 12.9 years. Thus, we assume that a purchased car will be used for a slightly longer period, 15 years. We could obtain the effect of CO_2 emissions for 15 years of vehicle substitution and fuel substitution, as summarized in column (7). Further, the total effect is calculated in column (8), by summing up the values of columns (5) and (7).

[12]We assume that there is no difference among the regions on the choice of hydrogen production technologies.

Table 8.2 Hydrogen production by SRM only

	(1) Gasoline vehicle purchases	(2) FCV purchases	(3) Gasoline purchases	(4) Hydrogen purchases	(5) Vehicle substitution (2–1)	(6) Fuel substitution (4–3)	(7) Total fuel substitution	(8) Total effect (5 + 7)	Unit
Final demand	197,140	197,140	5,878	3,545	0	−2,334	−35,004	−35,004	Mil. yen
Aichi	54,710	197,140	457	276	142,400	−182	−2,723	139,677	
ROJ	142,400	0	5,421	3,269	−142,400	−2,152	−32,281	−174,681	
Production	509,147	500,543	7,383	4,741	−8,604	−2,641	−39,622	−48,227	Mil. yen
Aichi	126,518	321,490	554	372	194,971	−182	−2,730	192,242	
ROJ	382,629	179,053	6,828	4,369	−203,576	−2,460	−36,893	−240,468	
Gross value added	165,829	157,931	2,788	3,288	−7,898	500	7,504	−394	Mil. yen
Aichi	39,485	92,172	207	255	52,686	48	719	53,405	
ROJ	126,344	65,760	2,581	3,033	−60,584	452	6,785	−53,799	
Employment	25,255	25,298	338	109	43	−229	−3,435	−3,392	Persons
Aichi	6,069	17,769	25	8	11,699	−17	−256	11,444	
ROJ	19,186	7,530	314	102	−11,656	−212	−3,180	−14,836	
CO_2 emissions	509,109	689,974	114,071	22,890	180,865	−91,181	−1,367,719	−1,186,854	t-CO_2
Aichi	99,807	257,223	8,853	1,793	157,416	−7,061	−105,912	51,504	
ROJ	409,302	432,750	105,218	21,098	23,449	−84,120	−1,261,807	−1,238,358	
Industry sector	509,109	689,974	14,999	22,890	180,865	7,891	118,361	299,226	t-CO_2
Aichi	99,807	257,223	1,146	1,793	154,416	646	9,694	167,110	
ROJ	409,302	432,750	13,853	21,098	23,449	7,244	108,667	132,116	
Household	0	0	99,072	0	0	−99,072	−1,486,080	−1,486,080	t-CO_2
Aichi	0	0	7,707	0	0	−7,707	−115,606	−115,606	
ROJ	0	0	91,365	0	0	−91,365	−1,370,474	−1,370,474	

Source: Authors' calculations

Considering the accumulated effect of fuel substitution, CO_2 emissions for both the vehicle substitution in the first year and the fuel substitution, which occurs when using the car for 15 years, increase by 180,865 t-CO_2 and 118,361 t-CO_2, respectively, for a total of 299,226 t-CO_2. The effect in the household sector is a reduction of 1,486,080 t-CO_2; the total effect is a decrease of 1,186,854 t-CO_2.[13]

Table 8.3 shows the simulation results when hydrogen is produced by DDM. The final demand in each case is the same as in Table 8.2, and the direction of the effects is almost the same. The difference appears in case 4, the hydrogen purchase, and then in cases 6 and 7, the fuel substitution. With respect to CO_2 emissions, the fuel substitution effect per year is a decrease of 9371 t-CO_2 in the industrial sector, because the hydrogen produced by DDM lowers t-CO_2 emissions, although SRM-produced hydrogen increases CO_2 emissions by 7891 t-CO_2. The total effects are CO_2 emissions decrease by 1,445,777 t-CO_2 (a 40,303 t-CO_2 increase in the industry sector and a 1,486,080 t-CO_2 decrease in the household sector). Even though the effect of the household sector remains dominant, it is possible to suppress the CO_2 increase in the industrial sector, when hydrogen is produced by DDM.

Hydrogen production only by SRM reduces the total CO_2 emissions by 1,186,854 t-CO_2, whereas hydrogen production only by DDM reduces them by 1,445,777 t-CO_2. The latter reduces CO_2 emissions by approximately 21.8% more than the former. The effect of vehicle substitution is also constant at 180,865 t-CO_2 regardless of the choice of hydrogen production technology. FCV production generates more CO_2 emissions than gasoline vehicle production. As fuel substitution in the industry, hydrogen production only by SRM increases CO_2 emissions by 118,361 t-CO_2. On the other hand, hydrogen production only by DDM reduces CO_2 emissions by 140,561 t-CO_2.

Tables 8.2 and 8.3 also show the regional effects, for Aichi Prefecture and the ROJ. Although gasoline vehicles are produced in both regions, FCVs are produced only in Aichi Prefecture. Therefore, the demand shift from gasoline vehicles to FCVs requires concentration of production in Aichi Prefecture, which increases the prefecture's production by 194,971 million yen and decreases the production of the ROJ by 203,576 million yen. CO_2 emissions increase by 154,416 t-CO_2 in Aichi Prefecture and by 23,449 t-CO_2 in ROJ. These effects are independent of the choice of the hydrogen production technologies SRM and DDM.

Purchased vehicles are assumed to be distributed proportional to the regional demand; therefore, the effects of fuel substitution have analogous regional tendencies. The directions of the effects are the same in all regions, although the sales volumes differ according to the regional demand. If the production in both regions decreases, then CO_2 emissions also decrease in both regions, although CO_2 emissions in the industry increase in both regions for SRM. The total effect of CO_2 emissions in the industry differs by regions and technology choice. With SRM, the CO_2 emissions in Aichi Prefecture increase by 167,110 t-CO_2, and those in the ROJ

[13]These values are interpreted as the stable states of the changes when a certain amount of increases in final demands are sustained.

Table 8.3 Hydrogen production by DDM only

	(1) Gasoline vehicle purchases	(2) FCV purchases	(3) Gasoline purchases	(4) Hydrogen purchases	(5) Vehicle substitution (2–1)	(6) Fuel substitution (4–3)	(7) Total fuel substitution	(8) Total effect (5 + 7)	Unit
Final demand	197,140	197,140	5,878	3,545	0	−2,334	−35,004	−35,004	Mil. yen
Aichi	54,710	197,140	457	276	142,400	−182	−2,723	139,677	
ROJ	142,400	0	5,421	3,269	−142,400	−2,152	−32,281	−174,681	
Production	509,147	500,543	7,383	5,652	−8,604	−1,730	−25,956	−34,561	Mil. yen
Aichi	126,518	321,490	554	459	194,971	−95	−1,426	193,546	
ROJ	382,629	179,053	6,828	5,193	−203,576	−1,635	−24,531	−228,107	
Gross value added	165,829	157,931	2,788	2,977	−7,898	189	2,839	−5,059	Mil. yen
Aichi	39,485	92,172	207	235	52,686	28	425	53,111	
ROJ	126,344	65,760	2,581	2,742	−60,584	161	2,414	−58,170	
Employment	25,255	25,298	338	136	43	−202	−3,034	−2,991	Persons
Aichi	6,069	17,769	25	10	11,699	−15	−220	11,479	
ROJ	19,186	7,530	314	126	−11,656	−188	−2,814	−14,470	
CO_2 emissions	509,109	689,974	114,071	5,629	180,865	−108,443	−1,626,641	−1,445,777	t-CO_2
Aichi	99,807	257,223	8,853	469	157,416	−8,384	−125,764	31,652	
ROJ	409,302	432,750	105,218	5,159	23,449	−100,059	−1,500,878	−1,477,429	
Industry sector	509,109	689,974	14,999	5,629	180,865	−9,371	−140,561	40,303	t-CO_2
Aichi	99,807	257,223	1,146	469	154,416	−677	−10,157	147,258	
ROJ	409,302	432,750	13,853	5,159	23,449	−8,694	−130,404	−106,955	
Household	0	0	99,072	0	0	−99,072	−1,486,080	−1,486,080	t-CO_2
Aichi	0	0	7,707	0	0	−7,707	−115,606	−115,606	
ROJ	0	0	91,365	0	0	−91,365	−1,370,474	−1,370,474	

Source: Authors' calculations

also increase by 132,116 t-CO_2. However, DDM raises CO_2 emissions in Aichi Prefecture by 147,258 t-CO_2 but lowers them by 106,955 t-CO_2 in the ROJ.

Overall, SRM increases CO_2 emissions by 51,504 t-CO_2 in Aichi Prefecture and reduces them by 1,238,358 t-CO_2 in the ROJ. However, when DDM technology is adopted, the increase of CO_2 emissions in Aichi Prefecture is suppressed to 31,652 t-CO_2, and their decrease in the ROJ much expands to 1,477,429 t-CO_2.

8.6 Concluding Remarks

In this study, we analyze the economic and environmental effects of defusing FCVs that use hydrogen fuel based on the selection of several production technologies. Overall, the effects on society, production, value added, employment, and CO_2 emissions are obtained by a scenario input-output analysis. With regard to the hydrogen production technology, we focused on comparing the steam reforming method, currently considered mainstream, and our newly developed methane direct decomposition method. The findings obtained are as follows:

1. Substituting conventional vehicles with FCVs decreases the production and value added, because the ripple effect of producing FCVs is relatively small compared to that of conventional vehicles. In spite of this, CO_2 emissions increase, because production of FCVs causes more CO_2 emissions than conventional vehicles.
2. Fuel substitution of hydrogen for gasoline has a total effect of reducing CO_2 emissions. The effect on the industrial sector depends on the choice of hydrogen production technology; the effect on the household sector is more substantial.
3. In the hydrogen production sector, both production technologies have direct CO_2 emissions; however, direct CO_2 emissions by DDM are 14.1% of those by SRM, which is much lower.
4. In addition, substitution of fuels by SRM in the industrial sector adds 118,361 t-CO_2, and that by DDM removes 140,561 t-CO_2. The total effect is a reduction in CO_2 emissions by 1,445,777 t-CO_2 for DDM and by 1,186,854 t-CO_2 for SRM. DDM reduces CO_2 emissions 21.8% more than SRM.
5. Regionally, vehicle substitution requires concentration of vehicle production in Aichi Prefecture, implying that the production, value added, and employment in that prefecture increase. Therefore, CO_2 emissions also increase there, despite the fuel substitution. However, DDM technology is able to suppress the increase in CO_2 emissions, although SRM augments them.

In our analysis, we do not consider the construction effect of a hydrogen refueling station or the capital investment effect due to the expansion of automobile production. These effects increase production, value added, employment, and CO_2 emissions and, therefore, reduce the effect evaluated in this study. Furthermore, we focused on the diffusion of FCVs and the selection of hydrogen production technology, but hydrogen use is not only limited to vehicles. There exists the possibility of using hydrogen for fuel combustion and generating electric power.

These issues concerning the economic and environmental effects of the technology choice will be the focus of our future research.

Acknowledgment This paper is part of the research results of "Development of the Methane direct decomposition hydrogen production system" of Project E, Priority Research Project II, at Knowledge Hub Aichi, Japan. We are grateful for the financial support.

References

Cantono S, Heijungs R, Kleijn R (2008) Environmental accounting of eco-innovations through environmental input-output analysis: the case of hydrogen and fuel cells buses. Econ Syst Res 20(3):303–318

Chubu region Institute for Social and Economic Research (2015) On the Influence of Next Generation Mobility Promotion in the Chubu Region Industry (in Japanese)

Fujikawa K, Wang J (2017) Economic and environmental impact of renewable energy – an application of input-output analysis. Ritsumeikan Econ Rev 65(4):619–630. (in Japanese)

Hydrogen and Fuel Cell Strategy Council, METI (2016) Hydrogen and Fuel Cell Strategy Roadmap – accelerate efforts toward Realization of a Hydrogen Society (in Japanese). http://www.meti.go.jp/press/2015/03/20160322009/20160322009.html

Japan Automobile Research Institute (2011) Report on overall efficiency and GHG emissions by type of vehicle (in Japanese). http://www.jari.or.jp/Portals/0/jhfc/data/report/2010/pdf/result.pdf

Keipi T, Tolvanen H, Konttinen J (2018) Economic analysis of hydrogen production by methane thermal decomposition: comparison to competing technologies. Energy Convers Manag 159:264–273

Leontief W (1970) Environmental repercussions and the economic structure: an input-output approach. Rev Econ Stat 52(3):262–271

Miller RE, Blair PD (2009) Input-output analysis: foundations and extensions, 2nd edn. Cambridge University Press, Cambridge

Ministry of Environment (2017) Emissions of greenhouse gases in Japan (preliminary figures for (2016)) (in Japanese). https://www.env.go.jp/press/104900.html

Nansai K, Kagawa S, Kondo Y, Suh S, Inaba R, Nakajima K (2009) Improving the completeness of product carbon footprints using a global link input-output model: the case of Japan. Econ Syst Res 21(3):267–290

National Institute for Environmental Studies (2018) Embodied Energy and Emission Intensity Data for Japan Using Input-Output tables (3EID). (in Japanese). http://www.cger.nies.go.jp/publications/report/d031/jpn/index_j.htm

Next Generation Vehicle Promotion Center (2018) Statistics of the next generation vehicle (in Japanese). http://www.cev-pc.or.jp/chosa/

Usubiaga A, Acosta-Fernandez J (2015) Carbon emission accounting in MRIO models: the territory vs. residence principle. Econ Syst Res 27(4):458–477

Wang J (2016) Economic and environmental effects of introduction of renewable power sources: an application of scenario input-output analysis to China. Input-Output Anal 24(1):35–48. (in Japanese)

Wiebe KS, Bruckner M, Giljum S, Lutz C (2012) Calculating energy-related CO_2 emissions embodied in international trade using global input-output model. Econ Syst Res 24(2):113–139

Yoshioka K, Suga M (1997) Applications of the input-output approach in environmental analysis – a study of scenario Leontief inverse, economic analysis. Economic and Social Research Institute, Cabinet Office, Japan, 154 (in Japanese)

Chapter 9
An Evaluation of Environmental Load Reduction in Mikawa Bay: The Input–Output Model Approach

Hiroyuki Shibusawa, Risa Ochiai, and Katsuhiro Sakurai

Abstract Mikawa Bay is a very shallow inner bay situated between the Atsumi and Chita peninsulas in the Aichi prefecture. The port section is not very large and is a closed ocean area with minimal external seawater exchange. This makes it easy for pollutants to build up, and the area is known for frequent outbreaks of red or discolored tides. Therefore, purifying the Mikawa Bay environment (i.e., infrastructure, sewage systems, septic tanks, private and industrial wastewater, seawater, and sediment pollution) has become a priority. The objective of this study is to provide a quantitative evaluation of the environmental impact of economic activities in the Mikawa Bay watershed and surrounding regions (e.g., Toyogawa, Yahagi, Sakaigawa river basins, Atsumi Bay watershed, and Chita Bay watershed). We estimate an interregional input–output table of the Aichi prefecture and analyze the relationship between each city and town, and their environmental impact on Mikawa Bay, considering economic trade relationships using the input–output and optimization models.

Keywords Mikawa Bay · Basin · Environmental emissions · Input–output model · Optimization model

9.1 Introduction

Watershed and water resource management are of the utmost importance for daily life and socioeconomic activity. This is reiterated by the Basic Law on the Water Cycle of Japan, enforced in July 2014. This Act was aimed at the maintenance or

---

H. Shibusawa (✉) · R. Ochiai
Department of Architecture and Civil Engineering, Toyohashi University of Technology, Toyohashi, Japan
e-mail: hiro-shibu@tut.jp

K. Sakurai
Faculty of Economics, Rissho University, Tokyo, Japan
e-mail: k-sakurai@ris.ac.jp

K. Nakayama, Y. Miyata (eds.), *Theoretical and Empirical Analysis in Environmental Economics*, New Frontiers in Regional Science: Asian Perspectives 34, https://doi.org/10.1007/978-981-13-2363-8_9

restoration of a sound water cycle in conjunction with the healthy development of Japan's economy. Water usage and water-based environments are closely tied to the characteristics of economic activity in a region or urban area.

Aichi prefecture, the focus of this study, is among the top prefectures in the nation with regard to manufacturing, particularly transport equipment. Commerce, agriculture, and fisheries are also prevalent, giving the prefecture a well-balanced industrial structure. Additionally, the Aichi prefecture has a diverse regional environment, with cities, farmland, nature, rivers, and ocean areas. Sustainable watershed management requires an examination of various measures for restoring or regenerating environmental deterioration. The industrial structure and geographical characteristics of the Aichi prefecture make it appropriate as a regional model for environmental symbiosis.

Mikawa Bay, a shallow inner bay, is situated between the Atsumi and Chita peninsulas in Aichi prefecture. The port section is not very large and is a closed ocean area with little exchange of outside sea water. This makes it easy for pollutants to accumulate, and the ocean area has frequent outbreaks of red or discolored tides. Therefore, purifying the Mikawa Bay (i.e., infrastructure, sewage systems, septic tanks, private and industrial wastewater, seawater, and sediment pollution) has become a priority.

The study region is an area comprised of 25 cities and towns in the Aichi prefecture of the Mikawa Bay area and along the primary rivers in the Mikawa watershed region, including 27 other regions therein. The focus of the study is environmentally hazardous substances that play an important role in either improving or worsening the water environment of Mikawa Bay. We estimate the amount of environmentally hazardous substances created by economic activities of the cities and towns in the eastern Owari, Chita, West Mikawa, and East Mikawa areas within the Mikawa watershed.

The objective is thus to quantitatively evaluate the environmental impact of economic activities of Nagoya and other cities and towns in the Toyokawa, Yahagi, Sakagawa, and Atsumi river basins that flow into Mikawa Bay. We create an interregional input–output (IRIO) table of the Mikawa Bay watershed region and analyze the relationship between each city and town and the environmental impact on Mikawa Bay by considering the economic trade relationships between these cities and towns.

Comprehensive surveys of the environmental impact on Mikawa Bay include Fukuda and Takayanagi (2009), which estimated the environmental impact units and outflow coefficients. There are examples of extant research of the Toyokawa watershed (Yamaguchi and Shibusawa 2000, 2005; Shibusawa and Yamaguchi 2014) and studies that have analyzed the effect of watershed management policies using dynamic optimization models (Sakurai et al. 2016; Sakurai et al. 2017). Other studies estimated environmental impact of the economies of cities, towns, and villages in the Toyokawa watershed (Ochiai et al. 2017). However, this specific study is a new attempt at an environmental economic analysis of the overall Mikawa watershed.

9.2 Research Methodology

For the purpose of this study, the Mikawa Bay watershed region is defined as the Toyokawa, Yahagi, Sakagawa, Atsumi, and Chita watersheds, as shown in Table 9.1 and Fig. 9.1 (Fukuda and Takayanagi 2009). Because the watershed areas and administrative units of the cities and towns do not strictly align, the portions of the cities and towns included in the watershed regions are calculated by the number of employees in each geographic unit according to an economic census. While the city of Nagoya and other parts of Aichi prefecture are outside the Mikawa Harbor area, because we are studying the entire Aichi prefecture, these two regions, with their large economies, are included.

We combine regions and sectors from a 2011 IRIO table for the Aichi prefecture. We then prepare an IRIO table for the Mikawa Bay watershed region. Using an

Table 9.1 Cities and towns, regions, and watersheds in the Mikawa Bay watershed region

	Municipality	Region	Sub-region	Basin
1	Nagoya City	East Owari	East Owari	–
2	Toyoake City			Sakai River
3	Togo Town			Sakai River
4	Handa City	Chita	Chita	Chita Bay
5	Obu City			Chita Bay
6	Agui Town			Chita Bay
7	Higashiura Town			Chita Bay
8	Minamichita Town			Chita Bay
9	Mihama Town			Chita Bay
10	Taketoyo Town			Chita Bay
11	Toyota City	West Mikawa	Toyota/Miyoshi	Yahagi River
12	Miyoshi City			Yahagi River
13	Okazaki City		Okazaki/Koda	Yahagi River
14	Koda Town			Atsumi Bay
15	Hekinan City		East Kinuura	Chita Bay
16	Kariya City			Chita Bay
17	Anjyo City			Chita Bay
18	Chiryu City			Sakai River
19	Takahama City			Chita Bay
20	Nishio City		Nishio	Yahagi River
21	Shinshiro City	East Mikawa	Shinshiro/Shitara	Toyogawa River
22	Shitara Town			Toyogawa River
23	Toyohashi City		Toyohashi/Atsumi	Atsumi Bay
24	Gamagori City			Atsumi Bay
25	Tahara City			Atsumi Bay
26	Toyokawa City		Toyokawa	Toyokawa River
27	Rest of Aichi			–

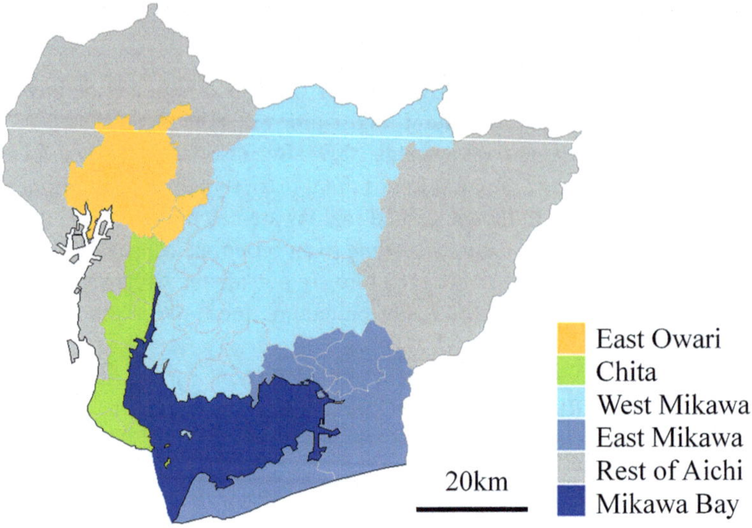

Fig. 9.1 The region

input–output model, we estimate the amount of production induced by the final demands of cities and towns, and we calculate environmental impact on Mikawa Bay using environmental impact units and an outflow ratio. We also create an optimization model that maximizes regional gross domestic product (GDP) of the Mikawa Bay watershed region using supply and demand and environmental conditions. We then examine the nature of industrial composition considering reduced environmental impact.

9.3 Analysis

9.3.1 Input–Output Model

We estimate an IRIO table using a 2011 Aichi prefecture input–output table (see Yamada and Owaki 2012). The integration of regions allows for the creation of an IRIO table with 25 cities and towns, the city of Nagoya in the Mikawa Bay watershed region, 27 regions, and 33 sectors in the Aichi prefecture. The production induced by demand from each village and town, as well as the environmental emissions (i.e., chemical oxygen demand (COD), total nitrogen (T-N) and total phosphorus (T-P)) flowing into the Mikawa Bay, was estimated, using an IRIO model.

In the context of this economy, n is the number of regions (e.g., cities and towns), and m is the number of industry sectors. The amount of induced production using a simple form is given by

$$\mathbf{X} = \left(\mathbf{I} - \left(\mathbf{I} - \widehat{\mathbf{M}}\right)\mathbf{A}\right)^{-1}\mathbf{F}, \tag{9.1}$$

where **X** is the production column vector; **A** is the input coefficient matrix; **I** is the unit matrix; $\widehat{\mathbf{M}}$ is the import rate matrix; and **F** is the final demand column vector. represents the diagonal matrix (see Miller and Blair 2009).

Because we estimate the environmental emissions flowing into the Mikawa Bay using production induced by the region in question and other regions according to the final demand of each city and town, we find the amount of induced production caused by the final demand. Postulating a Leontief inverse matrix as

$$\mathbf{B} = \left[\mathbf{I} - \left(\mathbf{I} - \widehat{\mathbf{M}}\right)\mathbf{A}\right]^{-1}, \tag{9.2}$$

the production induced by each city and town is deconstructed as follows.

$$\mathbf{X} = \mathbf{BF} = \mathbf{BF_1} + \mathbf{BF_2} + \cdots + \mathbf{BF_n}. \tag{9.3}$$

In this formula, $\mathbf{F_i}$ is the final demand column vector for city or town i (the column vector of total final demand for each city and town), and the amount of production induced by the final demand of each city or town, i, is $\mathbf{BF_i}$. Placing this in $\widetilde{\mathbf{X}}_\mathbf{i} = \mathbf{BF_i}$, and by creating rows for all cities and towns, we obtain the induced production matrix, $\widetilde{\mathbf{X}}$ (nm rows × n columns).

$$\widetilde{\mathbf{X}} = \left[\widetilde{\mathbf{X}}_\mathbf{1}, \widetilde{\mathbf{X}}_\mathbf{2}, \ldots, \widetilde{\mathbf{X}}_\mathbf{n}\right]. \tag{9.4}$$

If $\widetilde{\mathbf{X}}$ is aggregated for each sector, the induced production matrix for each sector, $\mathbf{X_{sec}}$ (m rows × n columns), provides the induced production matrix, $\mathbf{X_{reg}}$ (n rows × n columns), for each city and town.

Environmental impact is found by multiplying induced production by city and town and the sector by the regional share, inflow percentage, COD, T-N, and T-P impact units. We explain this using COD as an example. $\mathbf{c}_{Org}^{\mathbf{COD}}$ is the COD coefficient where the environmental load is generated at the region of origin. γ is the regional share. δ is the inflow percentage to Mikawa Bay. Then, the adjusted COD impact coefficients of industry j of city or town i are calculated as $c_{ij}^{COD} = c_{Oij}^{COD}\gamma_i\delta_i$. The impact of induced COD found from the final demand of city or town i is derived from the following formula.

$$\mathbf{E_i^{COD}} = \widehat{\mathbf{c}}^{\mathbf{COD}}\widetilde{\mathbf{X}}_\mathbf{i}. \tag{9.5}$$

By creating a matrix for all cities and towns, we can solve for the induced COD impact matrix, $\mathbf{E^{COD}}$. If we aggregate this by sector, we obtain the induced COD impact matrix for each sector, $\mathbf{E}_{\mathbf{sec}}^{\mathbf{COD}}$ (m rows × n columns), and aggregating this by each city and town yields the induced COD impact matrix, $\mathbf{E}_{\mathbf{reg}}^{\mathbf{COD}}$ (n rows × n columns).

Using three regions (cities and towns) as an example, the induced COD impact matrix, aggregated by city and town, is given in the following formula (COD descriptions are omitted):

$$\mathbf{E}_{\mathbf{reg}} = \begin{bmatrix} E_{11} & E_{12} & E_{13} \\ E_{21} & E_{22} & E_{23} \\ E_{31} & E_{32} & E_{33} \end{bmatrix}. \tag{9.6}$$

The diagonal element, E_{rr}, shows the environmental impact induced within the region of the area in question, owing to the final demand in region r. E_{rs} shows the environmental impact induced in region s from the final demand in region r. E_{rr} is the effect within the region of the area in question, and E_{rs} is the spillover effect on region s from region r.

Additionally, F_{ij} is the final demand of city or town i from economic activities in city or town j, and with E_{ij}^{COD} being the induced COD impact from induced production, the induced environmental impact α_{ij}^{COD} is as follows.

$$\alpha_{ij}^{COD} = E_{ij}^{COD} / F_{ij}. \tag{9.7}$$

The larger the value of this coefficient, the greater the environmental emissions generated.

9.3.2 *The Optimization Model*

In the "Future Estimated Population by Region in Japan," calculated by the National Institute of Population and Social Security Research, the labor population (ages 15–64) of the Aichi prefecture is predicted to decrease by about 10% between 2010 and 2040. Let us consider efficient resource allocation in an economy where the labor and capital (gross added value) of the region in question decreases with the additional constraint of environmental impact. This can be done by formulating an optimization problem that maximizes GDP (value added) based on supply and demand of goods and services and impacting emission constraints.

The optimization model for the economy is thus formulated as follows.

$$\max_{\{\mathbf{X}\}} (\mathbf{v})^t \, \mathbf{X}. \tag{9.8}$$

Subject to

$$\left(\mathbf{I}-\widehat{\mathbf{M}}\right)\mathbf{F_L}+\mathbf{EX_L}\leq\left(\mathbf{I}-\left(\mathbf{I}-\widehat{\mathbf{M}}\right)\mathbf{A}\right)\ \mathbf{X}\leq\left(\mathbf{I}-\widehat{\mathbf{M}}\right)\mathbf{F_U}+\mathbf{EX_U} \tag{9.9}$$

$$\overline{\mathbf{k}}_{\mathrm{L}}\leq\widehat{\mathbf{v}}\mathbf{X}\leq\overline{\mathbf{k}}_{\mathrm{U}} \tag{9.10}$$

$$(\mathbf{c_{COD}})^t\ \mathbf{X}\leq\overline{e}_{\mathrm{COD}} \tag{9.11}$$

$$(\mathbf{c_{T-N}})^t\ \mathbf{X}\leq\overline{e}_{T-N} \tag{9.12}$$

$$(\mathbf{c_{T-P}})^t\ \mathbf{X}\leq\overline{e}_{T-P} \tag{9.13}$$

$$\mathbf{X}\geq\mathbf{0}, \tag{9.14}$$

in which:

$\mathbf{X}=(X_1,\ \ldots,X_Z)^t$: the production column vector (endogenous)
$\mathbf{F_U}=(F_{U1},\ \ldots,F_{Uz})^t$: the upper limit of the final demand column vector
$\mathbf{F_L}=(F_{L1},\ \ldots,F_{Lz})^t$: the lower limit of the final demand column vector
$\mathbf{EX_U}=(EX_{U1},\ \ldots,EX_{Uz})^{\mathrm{t}}$: the upper limit of the export column vector
$\mathbf{EX_L}=(EX_{L1},\ \ldots,EX_{Lz})^{\mathrm{t}}$: the lower limit of the export column vector
$\mathbf{A}=(a_{ij})$: input coefficient matrix
$\mathbf{M}=(m_1,\ \ldots,m_z)^t$: the import rate
$\widehat{\mathbf{M}}$: diagonal matrix of the import rate
$\mathbf{v}=(v_1,\ \ldots,v_z)^t$: the rate of gross added value
$\widehat{\mathbf{v}}$: diagonal matrix of the rate of gross added value
$\overline{\mathbf{k}}_{\mathrm{U}}=\left(\overline{k}_{U1},\ldots,\overline{k}_{Uz}\right)^t$: the upper limit of supply of production factors
$\overline{\mathbf{k}}_{\mathrm{L}}=\left(\overline{k}_{L1},\ldots,\overline{k}_{Lz}\right)^t$: the lower limit of supply of production factors
$\mathbf{c_{COD}}=(c_{BOD1},\ \ldots,c_{BODz})^t$: the coefficient of COD emission
$\mathbf{c_{T-N}}=(c_{T-N1}\ \ldots,c_{T-Nz})^t$: the coefficient of T-N emission
$\mathbf{c_{T-P}}=(c_{T-P1}\ \ldots,c_{T-Pz})^t$: the coefficient of T-P emission
$\overline{e}_{COD}$: the constraint of the amount of COD pollutant inflow
$\overline{e}_{T-N}$: the constraint of the amount of T-N pollutant inflow
$\overline{e}_{T-P}$: the constraint of the amount of T-P pollutant inflow

From the table of 27 regions and 33 sectors, we combine regions and sectors to create an IRIO table of 10 regions and 18 sectors (i.e., $z=n\times m=10\times 18$). In our optimization model, the minimum constraints of labor and capital, and the minimum demand in supply and demand conditions for each city, town, and sector, are set to 80%. By adding total inflow constraints to the environmental emissions, the amount of production in each city and town decreases. However, our analysis is conducted assuming long-term conditions where, in certain regions and sectors, the labor population and capital utilization is allowed to drop to 20%.

9.4 Results of Analysis

9.4.1 Estimation of Induced Environmental Impact by City and Town

Figure 9.2 shows the amount of final demand and induced production for each city and town. Within the watershed area, demand and production from Toyota City, Toyohashi City, and Okazaki City are significant. Simultaneously, demand and production for Nagoya City and other areas in the Aichi prefecture are even greater, having a large percentage within the prefecture.

Figure 9.3 shows the induced environmental inflow to Mikawa Bay from each city and town. The COD, T-N, and T-P impact, induced by final demand in each city and town, are large for Anjyo City, Kariya City, Handa City, Hekinan City, and Toyokawa City. The demand and amount of production in these cities is not significant, but they are located near the Mikawa Bay. Owing to the relatively large inflow rate, they have higher induced environmental impact. The spatial COD inflow impacts are shown in Fig. 9.4.

Figure 9.5 shows the induced environmental impact for each sector. Aggregated by sector, food and beverage products have the highest value, followed by steel, chemical products, and transport equipment. Overall, the impact of T-N is greatest, followed by COD and T-P.

Table 9.2 shows the induced COD impact, $\mathbf{E}_{\mathbf{reg}}^{\mathbf{COD}}$ (kg/day). The table-side is the region where the demand rises, and the table-head shows the regions where environmental impact is induced. As the amount of inducement rises, the color shifts from green to yellow to red. The diagonal elements show the induced impact for each region in the area of question, with everything else being the spillover

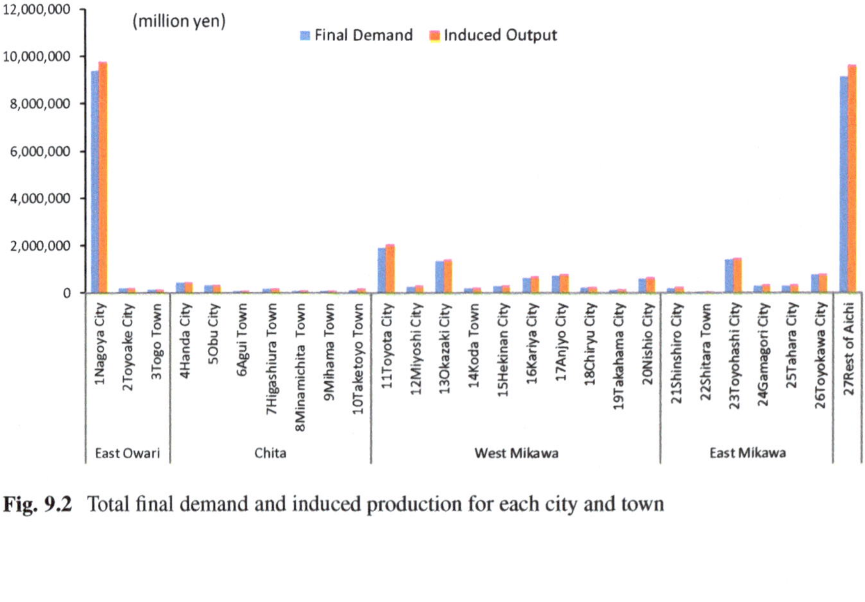

Fig. 9.2 Total final demand and induced production for each city and town

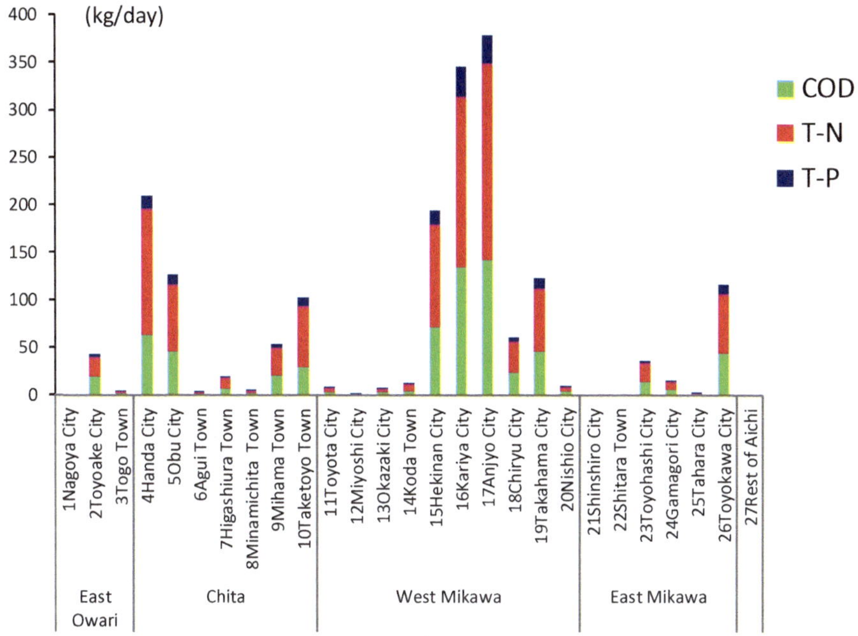

Fig. 9.3 Induced environmental impact for each city and town

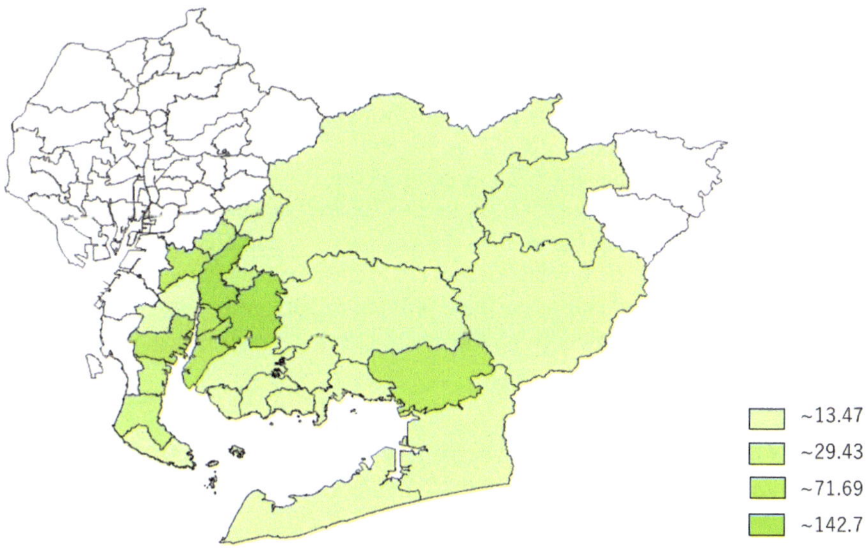

Fig. 9.4 Spatial impact of COD inflow

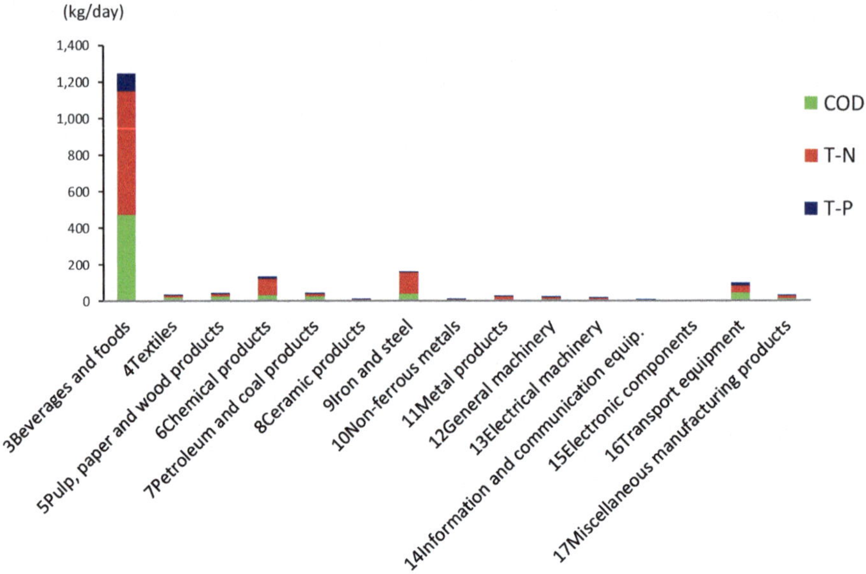

Fig. 9.5 Induced environmental impact by sector. (**a**) Region. (**b**) Sector

impact to other regions. For example, demand in Nagoya City does not induce any environmental impact within Nagoya City, which is not part of the watershed. However, it does induce a COD impact of 14.6 kg/day in Kariya City, 13.9 kg/day in Anjyo City, and 8.0 kg/day in Handa City. Cities within the area of question with higher induced impact are shown in red, including Kariya City, Anjyo City, Hekinan City, Handa City, Takahama City, and Toyokawa City.

Table 9.3 shows the induced COD coefficient, α_{ij}^{COD} (kg/day, millions of yen), with the amount of induced COD per million yen of demand. High inducement coefficients are shown in red. Regardless of being regions with high demand created, Taketoyo City, Mihama City, Takahama City, and Hekinan City tend to have high inducement coefficients. Compared to final demand in regions in the area of question, these cities and towns have a higher COD outflow to Mikawa Bay, owing to the spillover of induced production.

9.4.2 Optimization Model Accounting for Environmental Impact Constraints

The next step is to estimate the decrease in regional GDP when the constraints of COD, T-P, and T-N inflow are changed from 0% to 18% and are decreased in increments of 2%. The *x* axis shows the decrease of inflow impact, whereas the *y* axis shows the rate of change in regional GDP.

Table 9.2 Induced COD impact (kg/day)

Region where the demand arises		Regions where environmental impact is induced																											
		East Owari			Chita							West Mikawa										East Mikawa							
		1Nagoya City	2Toyoake City	3Togo Town	4Handa City	5Obu City	6Agui Town	7Higashiura Town	8Minamichita Town	9Mihama Town	10Taketoyo Town	11Toyota City	12Miyoshi City	13Okazaki City	14Koda Town	15Hekinan City	16Kariya City	17Anjyo City	18Chiryu City	19Takahama City	20Nishio City	21Shinshiro City	22Shitara Town	23Toyohashi City	24Gamagori City	25Tahara City	26Toyokawa City	27Rest of Aichi	Total
East Owari	1Nagoya City	0.0	2.8	0.2	8.0	6.9	0.1	1.1	0.3	2.5	3.6	0.5	0.1	0.2	0.3	6.9	14.6	13.9	1.4	4.1	0.5	0.0	0.0	0.9	0.5	0.2	3.0	0.0	72.6
	2Toyoake City	0.0	8.3	0.0	0.2	0.5	0.0	0.0	0.0	0.1	0.1	0.0	0.0	0.0	0.0	0.3	1.6	0.9	0.2	0.2	0.0	0.0	0.0	0.0	0.0	0.0	0.1	0.0	12.8
	3Togo Town	0.0	0.2	0.8	0.2	0.2	0.0	0.0	0.0	0.1	0.1	0.0	0.0	0.0	0.0	0.2	0.5	0.5	0.1	0.1	0.0	0.0	0.0	0.0	0.0	0.0	0.1	0.0	3.1
Chita	4Handa City	0.0	0.2	0.0	26.3	0.6	0.1	0.2	0.1	0.5	1.2	0.0	0.0	0.0	0.0	2.5	1.4	2.0	0.2	1.0	0.1	0.0	0.0	0.1	0.0	0.0	0.3	0.0	36.7
	5Obu City	0.0	0.4	0.0	0.5	21.0	0.0	0.1	0.0	0.1	0.2	0.0	0.0	0.0	0.0	0.6	2.3	1.3	0.2	0.4	0.0	0.0	0.0	0.0	0.0	0.0	0.1	0.0	27.4
	6Agui Town	0.0	0.0	0.0	0.8	0.1	0.8	0.0	0.0	0.1	0.1	0.0	0.0	0.0	0.0	0.3	0.3	0.4	0.0	0.2	0.0	0.0	0.0	0.0	0.0	0.0	0.0	0.0	3.3
	7Higashiura Town	0.0	0.1	0.0	0.6	0.5	0.0	2.8	0.0	0.1	0.2	0.0	0.0	0.0	0.0	0.5	1.1	1.0	0.1	0.5	0.0	0.0	0.0	0.0	0.0	0.0	0.1	0.0	7.7
	8Minamichita Town	0.0	0.0	0.0	0.1	0.1	0.0	0.0	0.7	0.4	0.2	0.0	0.0	0.0	0.0	0.2	0.2	0.2	0.0	0.1	0.0	0.0	0.0	0.0	0.0	0.0	0.1	0.0	2.4
	9Mihama Town	0.0	0.0	0.0	0.2	0.1	0.0	0.0	0.1	8.5	0.3	0.0	0.0	0.0	0.0	0.3	0.2	0.3	0.0	0.1	0.0	0.0	0.0	0.0	0.0	0.0	0.1	0.0	10.1
	10Taketoyo Town	0.0	0.1	0.0	1.0	0.1	0.0	0.0	0.0	0.5	10.9	0.0	0.0	0.0	0.0	0.7	0.4	0.6	0.0	0.2	0.0	0.0	0.0	0.0	0.0	0.0	0.1	0.0	14.9
West Mikawa	11Toyota City	0.0	0.8	0.1	2.1	1.4	0.0	0.2	0.1	0.7	1.2	1.4	0.0	0.2	0.1	2.0	4.1	5.2	0.5	1.0	0.2	0.0	0.0	0.4	0.2	0.0	1.7	0.0	23.7
	12Miyoshi City	0.0	0.4	0.3	0.4	0.3	0.0	0.1	0.0	0.1	0.2	0.0	0.2	0.0	0.0	0.3	1.1	0.9	0.1	0.2	0.0	0.0	0.0	0.0	0.0	0.0	0.2	0.0	5.0
	13Okazaki City	0.0	0.5	0.0	1.6	1.0	0.0	0.2	0.1	0.7	1.0	0.2	0.0	2.0	0.2	2.1	3.5	6.4	0.5	1.0	0.2	0.0	0.0	0.5	0.4	0.0	2.8	0.0	24.7
	14Koda Town	0.0	0.1	0.0	0.3	0.1	0.0	0.0	0.0	0.1	0.2	0.0	0.0	0.0	2.3	0.4	0.4	0.9	0.1	0.2	0.1	0.0	0.0	0.1	0.1	0.0	0.2	0.0	5.6
	15Hekinan City	0.0	0.1	0.0	1.8	0.3	0.0	0.1	0.0	0.3	0.6	0.0	0.0	0.0	0.0	34.0	1.0	1.8	0.1	0.8	0.2	0.0	0.0	0.1	0.0	0.0	0.2	0.0	41.5
	16Kariya City	0.0	1.2	0.0	1.2	1.7	0.0	0.2	0.0	0.3	0.5	0.0	0.0	0.0	0.0	1.2	71.3	4.2	1.8	1.0	0.1	0.0	0.0	0.1	0.1	0.0	0.3	0.0	85.4
	17Anjyo City	0.0	0.5	0.0	1.5	0.9	0.0	0.2	0.1	0.4	0.8	0.0	0.0	0.1	0.1	2.2	4.4	71.3	0.8	1.6	0.2	0.0	0.0	0.1	0.1	0.0	0.5	0.0	85.6
	18Chiryu City	0.0	0.3	0.0	0.3	0.4	0.0	0.0	0.0	0.1	0.2	0.0	0.0	0.0	0.0	0.5	4.4	2.1	15.0	0.3	0.0	0.0	0.0	0.0	0.0	0.0	0.1	0.0	23.9
	19Takahama City	0.0	0.1	0.0	0.6	0.2	0.0	0.1	0.0	0.1	0.2	0.0	0.0	0.0	0.0	0.9	0.9	1.4	0.1	26.0	0.0	0.0	0.0	0.0	0.0	0.0	0.1	0.0	30.7
	20Nishio City	0.0	0.3	0.0	1.9	0.6	0.0	0.1	0.1	0.7	1.0	0.0	0.0	0.0	0.1	4.4	2.1	5.1	0.3	1.1	1.6	0.0	0.0	0.2	0.1	0.0	0.5	0.0	20.4
East Mikawa	21Shinshiro City	0.0	0.1	0.0	0.3	0.1	0.0	0.0	0.0	0.1	0.2	0.0	0.0	0.0	0.0	0.2	0.4	0.5	0.0	0.1	0.0	0.1	0.0	0.1	0.0	0.0	0.5	0.0	3.0
	22Shitara Town	0.0	0.0	0.0	0.1	0.0	0.0	0.0	0.0	0.0	0.0	0.0	0.0	0.0	0.0	0.0	0.1	0.1	0.0	0.0	0.0	0.0	0.0	0.0	0.0	0.0	0.0	0.0	0.5
	23Toyohashi City	0.0	0.3	0.0	1.4	0.7	0.0	0.1	0.1	0.6	1.0	0.1	0.0	0.1	0.1	1.5	2.1	3.1	0.2	0.7	0.1	0.0	0.0	8.6	0.3	0.0	3.9	0.0	25.2
	24Gamagori City	0.0	0.1	0.0	0.4	0.2	0.0	0.0	0.0	0.2	0.3	0.0	0.0	0.0	0.1	0.5	0.6	1.2	0.1	0.2	0.0	0.0	0.0	0.2	3.3	0.0	0.7	0.0	8.2
	25Tahara City	0.0	0.1	0.0	0.5	0.2	0.0	0.0	0.0	0.3	0.3	0.0	0.0	0.0	0.0	0.5	0.6	0.8	0.1	0.2	0.1	0.0	0.0	0.2	0.1	0.3	0.4	0.0	5.0
	26Toyokawa City	0.0	0.2	0.0	0.9	0.4	0.0	0.1	0.0	0.3	0.5	0.1	0.0	0.1	0.1	0.8	1.3	1.9	0.1	0.4	0.1	0.0	0.0	0.7	0.2	0.0	24.3	0.0	32.6
	27Rest of Aichi	0.0	2.7	0.3	10.2	7.3	0.1	1.1	0.4	2.8	4.5	0.5	0.1	0.3	0.3	7.6	14.3	14.7	1.4	4.3	0.5	0.0	0.0	1.0	0.5	0.1	3.2	0.0	78.2
	Total	0.0	19.9	1.9	63.2	46.1	1.3	7.0	2.3	20.7	29.4	3.2	0.5	3.3	4.1	71.7	135.1	142.7	23.2	46.0	4.2	0.2	0.0	13.5	6.1	0.9	43.7	0.0	690.5

Table 9.3 Induced COD coefficient (kg/day in millions of yen)

Region where the demand arises			Regions where environmental impact is induced																										
			East Owari			Chita							West Mikawa										East Mikawa							
			1Nagoya City	2Toyoake City	3Togo Town	4Handa City	5Obu City	6Agui Town	7Higashiura Town	8Minamichita Town	9Mihama Town	10Taketoyo Town	11Toyota City	12Miyoshi City	13Okazaki City	14Koda Town	15Hekinan City	16Kariya City	17Anjyo City	18Chiryu City	19Takahama City	20Nishio City	21Shinshiro City	22Shitara Town	23Toyohashi City	24Gamagori City	25Tahara City	26Toyokawa City	27Rest of Aichi	Total
Region where the demand arises	East Owari	1Nagoya City	0.0	0.1	0.0	0.4	0.2	0.0	0.1	0.1	0.6	0.8	0.0	0.0	0.0	0.1	0.5	0.3	0.3	0.2	0.5	0.0	0.0	0.0	0.0	0.0	0.0	0.1	0.0	0.0
		2Toyoake City	0.0	0.1	0.0	0.2	0.1	0.0	0.0	0.1	0.5	0.6	0.0	0.0	0.0	0.0	0.5	0.1	0.3	0.1	0.4	0.0	0.0	0.0	0.0	0.0	0.0	0.1	0.0	0.1
		3Togo Town	0.0	0.1	0.0	0.3	0.2	0.0	0.1	0.1	0.6	0.6	0.0	0.0	0.0	0.0	0.5	0.2	0.3	0.1	0.4	0.0	0.0	0.0	0.0	0.0	0.0	0.1	0.0	0.0
	Chita	4Handa City	0.0	0.1	0.0	0.1	0.1	0.0	0.0	0.0	0.2	0.2	0.0	0.0	0.0	0.0	0.2	0.1	0.2	0.1	0.2	0.0	0.0	0.0	0.0	0.0	0.0	0.1	0.0	0.1
		5Obu City	0.0	0.0	0.0	0.1	0.2	0.0	0.0	0.1	0.3	0.5	0.0	0.0	0.0	0.0	0.4	0.1	0.2	0.1	0.3	0.0	0.0	0.0	0.0	0.0	0.0	0.1	0.0	0.1
		6Agui Town	0.0	0.1	0.0	0.1	0.1	0.0	0.0	0.0	0.2	0.3	0.0	0.0	0.0	0.0	0.3	0.1	0.2	0.1	0.2	0.0	0.0	0.0	0.0	0.0	0.0	0.1	0.0	0.0
		7Higashiura Town	0.0	0.1	0.0	0.1	0.1	0.0	0.0	0.0	0.3	0.3	0.0	0.0	0.0	0.0	0.3	0.1	0.2	0.1	0.2	0.0	0.0	0.0	0.0	0.0	0.0	0.1	0.0	0.0
		8Minamichita Town	0.0	0.1	0.0	0.1	0.2	0.0	0.1	0.0	0.1	0.2	0.0	0.0	0.0	0.0	0.3	0.2	0.2	0.1	0.3	0.0	0.0	0.0	0.0	0.0	0.0	0.1	0.0	0.0
		9Mihama Town	0.0	0.1	0.0	0.1	0.2	0.0	0.1	0.0	0.2	0.1	0.0	0.0	0.0	0.0	0.3	0.2	0.2	0.1	0.3	0.0	0.0	0.0	0.0	0.0	0.0	0.1	0.0	0.1
		10Taketoyo Town	0.0	0.1	0.0	0.1	0.1	0.0	0.0	0.0	0.1	0.2	0.0	0.0	0.0	0.0	0.2	0.2	0.2	0.1	0.3	0.0	0.0	0.0	0.0	0.0	0.0	0.1	0.0	0.1
	West Mikawa	11Toyota City	0.0	0.1	0.0	0.3	0.2	0.0	0.1	0.0	0.3	0.7	0.0	0.0	0.0	0.0	0.4	0.2	0.2	0.1	0.4	0.0	0.0	0.0	0.0	0.0	0.0	0.1	0.0	0.0
		12Miyoshi City	0.0	0.1	0.0	0.3	0.1	0.0	0.1	0.1	0.5	0.7	0.0	0.0	0.0	0.0	0.4	0.1	0.2	0.1	0.4	0.0	0.0	0.0	0.0	0.0	0.0	0.1	0.0	0.0
		13Okazaki City	0.0	0.1	0.0	0.2	0.2	0.0	0.1	0.0	0.3	0.5	0.0	0.0	0.0	0.0	0.4	0.2	0.2	0.1	0.3	0.0	0.0	0.0	0.0	0.0	0.0	0.0	0.0	0.0
		14Koda Town	0.0	0.1	0.0	0.2	0.2	0.0	0.0	0.0	0.2	0.5	0.0	0.0	0.0	0.0	0.2	0.1	0.1	0.1	0.3	0.0	0.0	0.0	0.0	0.0	0.0	0.0	0.0	0.0
		15Hekinan City	0.0	0.1	0.0	0.1	0.1	0.0	0.0	0.0	0.2	0.2	0.0	0.0	0.0	0.0	0.3	0.1	0.1	0.1	0.2	0.0	0.0	0.0	0.0	0.0	0.0	0.1	0.0	0.1
		16Kariya City	0.0	0.1	0.0	0.2	0.1	0.0	0.0	0.0	0.4	0.5	0.0	0.0	0.0	0.0	0.3	0.3	0.1	0.0	0.2	0.0	0.0	0.0	0.0	0.0	0.0	0.1	0.0	0.1
		17Anjyo City	0.0	0.1	0.0	0.1	0.1	0.0	0.0	0.0	0.2	0.4	0.0	0.0	0.0	0.0	0.2	0.1	0.2	0.0	0.2	0.0	0.0	0.0	0.0	0.0	0.0	0.1	0.0	0.1
		18Chiryu City	0.0	0.1	0.0	0.2	0.1	0.0	0.0	0.1	0.4	0.6	0.0	0.0	0.0	0.0	0.4	0.1	0.2	0.2	0.3	0.0	0.0	0.0	0.0	0.0	0.0	0.1	0.0	0.1
		19Takahama City	0.0	0.1	0.0	0.1	0.1	0.0	0.0	0.0	0.3	0.3	0.0	0.0	0.0	0.0	0.2	0.1	0.1	0.1	0.4	0.0	0.0	0.0	0.0	0.0	0.0	0.1	0.0	0.2
		20Nishio City	0.0	0.1	0.0	0.1	0.1	0.0	0.0	0.0	0.2	0.3	0.0	0.0	0.0	0.0	0.1	0.1	0.1	0.1	0.2	0.0	0.0	0.0	0.0	0.0	0.0	0.1	0.0	0.0
	East Mikawa	21Shinshiro City	0.0	0.1	0.0	0.3	0.2	0.0	0.1	0.0	0.3	0.7	0.0	0.0	0.0	0.0	0.4	0.2	0.2	0.1	0.4	0.0	0.0	0.0	0.0	0.0	0.0	0.0	0.0	0.0
		22Shitara Town	0.0	0.1	0.0	0.3	0.2	0.0	0.1	0.0	0.2	0.9	0.0	0.0	0.0	0.0	0.3	0.2	0.2	0.1	0.3	0.0	0.0	0.0	0.0	0.0	0.0	0.0	0.0	0.0
		23Toyohashi City	0.0	0.1	0.0	0.3	0.2	0.0	0.1	0.0	0.3	0.7	0.0	0.0	0.0	0.0	0.4	0.2	0.3	0.1	0.4	0.0	0.0	0.0	0.0	0.0	0.0	0.0	0.0	0.0
		24Gamagori City	0.0	0.1	0.0	0.2	0.2	0.0	0.1	0.0	0.2	0.5	0.0	0.0	0.0	0.0	0.4	0.2	0.2	0.1	0.4	0.0	0.0	0.0	0.0	0.0	0.0	0.0	0.0	0.0
		25Tahara City	0.0	0.1	0.0	0.2	0.2	0.0	0.1	0.0	0.1	0.4	0.0	0.0	0.0	0.0	0.3	0.2	0.2	0.1	0.3	0.0	0.0	0.0	0.0	0.0	0.0	0.0	0.0	0.0
		26Toyokawa City	0.0	0.1	0.0	0.3	0.2	0.0	0.1	0.0	0.3	0.7	0.0	0.0	0.0	0.0	0.4	0.2	0.2	0.1	0.4	0.0	0.0	0.0	0.0	0.0	0.0	0.1	0.0	0.0
		27Rest of Aichi	0.0	0.1	0.0	0.2	0.1	0.0	0.1	0.1	0.4	0.4	0.0	0.0	0.0	0.0	0.5	0.2	0.3	0.1	0.4	0.0	0.0	0.0	0.0	0.0	0.0	0.1	0.0	0.0
		Total	0.0	0.1	0.0	0.1	0.2	0.0	0.1	0.0	0.2	0.3	0.0	0.0	0.0	0.0	0.3	0.2	0.2	0.1	0.4	0.0	0.0	0.0	0.0	0.0	0.0	0.1	0.0	0.0

In terms of the rate of change in GDP by region (Fig. 9.6a) under the COD inflow constraints, the GDPs of Chita and west Mikawa and east Kinuura reduced markedly. The GDPs of east Mikawa (Toyokawa), west Mikawa (Nishio), east Owari, and east Mikawa (Toyohashi and Atsumi) dropped significantly when the level of COD decrease was 16% and below. GDP change by sector (Fig. 9.6b)

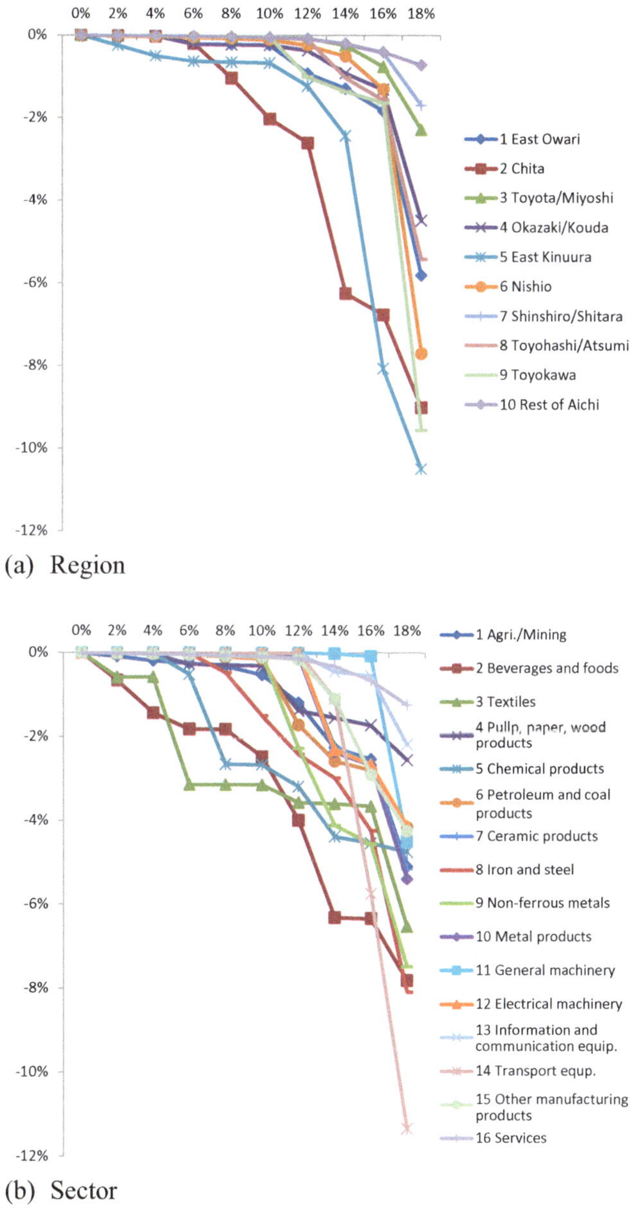

Fig. 9.6 Change in GDP due to COD inflow constraints. (**a**) Region. (**b**) Sector

observed a sharp drop in the GDPs (i.e., food and beverage products, textile products, and transport equipment).

Regarding regional GDP rate of change under T-N inflow constraints (Fig. 9.7a), the GDPs of Chita and west Mikawa (Kinuura-higashi) showed a significant drop. The GDPs of east Mikawa (Toyokawa) and west Mikawa (Nishio) also dropped significantly, beginning with a T-N decrease of 16%. Regarding the rate of change in GDP by sector (Fig. 9.7b), the change in chemical products, food, and beverage products was greatest up to a T-N decrease of 10%, after which the GDP decrease in steel became larger.

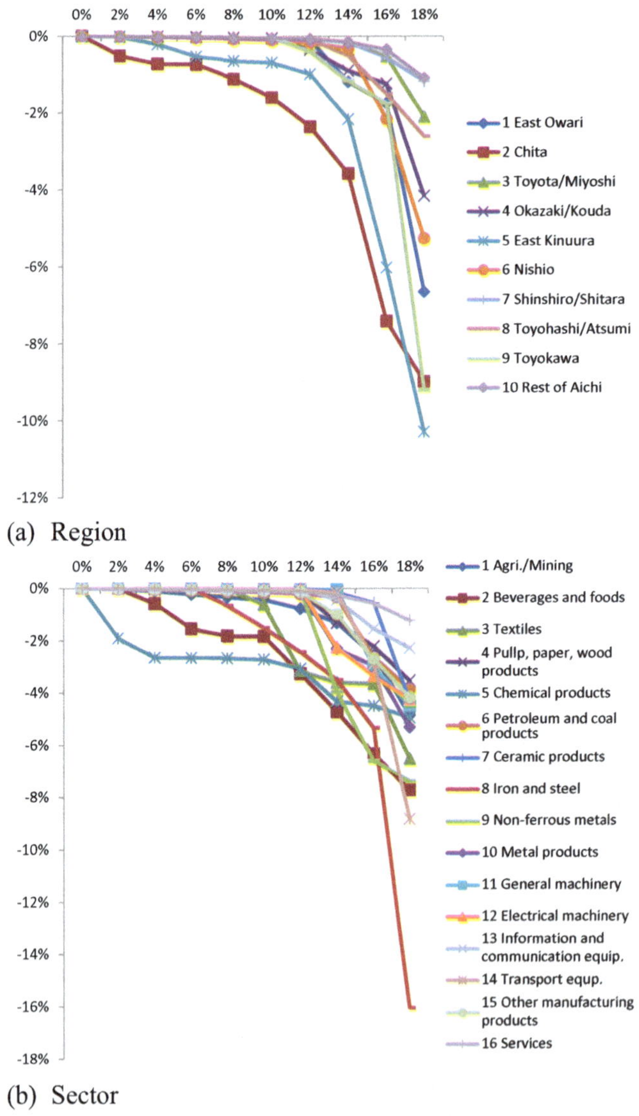

Fig. 9.7 Change in GDP due to T-N inflow constraints. (**a**) Region. (**b**) Sector

For regional GDP rate of change under T-P inflow constraints (Fig. 9.8a), the change in GDPs for Chita and west Mikawa (east Kinuura) was large. The changes in GDPs for east Mikawa (Toyokawa), west Mikawa (Nishio), eastern Owari, and east Mikawa (Toyohashi and Atsumi) grew, beginning with a T-P decrease of 16%.

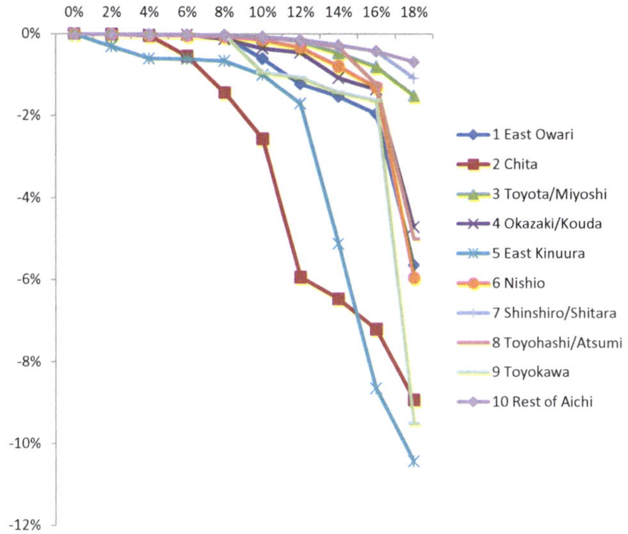

(a) Region

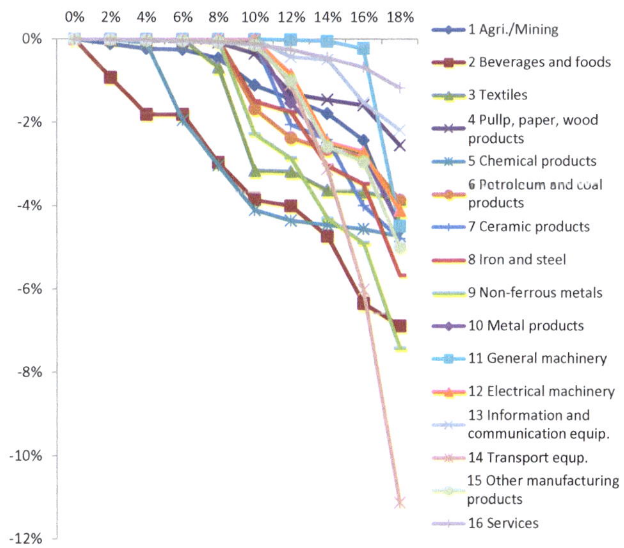

(b) Sector

Fig. 9.8 Change in GDP due to T-P inflow constraints. (**a**) Region. (**b**) Sector

Regarding the change in GDP by sector (Fig. 9.8b), the change was largest in food and beverage products, nonferrous metals, and transport equipment.

9.5 Conclusion

This study proposed an approach for evaluating the relationship between regional production activities and the environmental load at the municipal level in Mikawa Bay using input–output and optimization models.

The Mikawa Bay watershed has an agglomeration of industries, particularly focused on the transport equipment sector. The west Mikawa region, including Toyota City, is the site of numerous automotive factories. A large amount of environmental impact was thus predicted from that region, based on the transport industry. However, Toyota City is located inland and has a low outflow rate. We found no significant results with regard to the amount of inflow emissions.

In our estimate of environmental impact induced by demand using our input–output model, inflows from Anjyo City, Kariya City, Handa City, Hekinan City, and Toyokawa City, located near the Mikawa Bay, were significant. This is because these regions have a high outflow coefficient. We verified that the induced environmental impact was not only directly influenced by emissions from production activities, but also strongly influenced by geographical factors, resulting in the inflow to Mikawa Bay.

With our optimization model, we hypothesized a decrease in labor population, meaning a decrease in gross added value, within the area in question. We examined those regions and sectors which should have their production processes prioritized to decrease the inflow of emissions into Mikawa Bay while still maximizing regional GDP. An examination by region showed that production in Chita, west Mikawa (east Kinuura), and east Mikawa (Toyokawa), with west Mikawa (Nishio) and eastern Owari, had markedly decreased. Food, beverage products, and transport equipment tended to have large decreases. In the case of decreased T-N, steel had a large decrease in GDP. In the case of decreased T-P, nonferrous metals had a large decrease in GDP. The results of this analysis suggest that it is ideal to reduce production in these regions and sectors when prioritizing the maximization of GDP in the overall region.

A topic for future study is the clarification of how demand from outside Aichi prefecture (exports) affects the environmental impact on Mikawa Bay. One might also expand the model to evaluate economic and industrial subsidization policies or to consider changes in land usage via a spatially applied general equilibrium model. It has lastly been noted that a certain level of environmental impact on Mikawa Bay is necessary to maintain the production of clams and seaweed, which would thus also require an analysis from the perspective of the fisheries industry.

Acknowledgment This work was supported by JSPS KAKENHI Grant Number 17H01931, 17 K03726, and 17H02521.

References

Fukuda S, Takayanagi J (2009) Behavior material cycles in river basins and its influence on the environment of the semi-closed water area surrounded by the basins. Project research report of National Institute for Land and Infrastructure Management, No. 28 (in Japanese)

Hosei University, Japan Statistical Research Institute (2016) Municipality input-output table. https://www.hosei.ac.jp/toukei/index-j.html

Affluent Sea and Mikawa environment Regeneration Promotion Council (2017) http://www.mikawa-wan.jp/environment.html. Accessed 18 Sept 2017 (in Japanese)

Miller RE, Blair PD (2009) Input-output analysis, 2nd edn. Cambridge University Press, Cambridge

Ochiai R, Shibusawa H, Miyata Y, Sakurai K (2017) Evaluating the regional economy and the water environment in Mikawa Bay-Basin region, the 54th annual meeting of the JSRSAI, Kyoto, pp 1–6

Sakurai K, Shibusawa H, Nakayama K (2016) In: Kenkyu K (ed) The impacts of the enviromental and economic policy of the Toyogawa Basin, studies in port, vol 38. The Japan Port Economics Association, Chubu Branch, pp 21–38

Sakurai K, Shibusawa H, Mitsuhashi K, Kobayashi S (2017) Evaluation of the water-environment policy in the Toyogawa Basin, Japan. In: Shibusawa, Sakurai et al (eds) Socio-economic-environmental policies and evaluation in regional science. Springer, Cham, pp 651–666

Shibusawa H, Yamaguchi M (2014) Evaluating the social and economic impacts of the water resources in the Toyogawa Basin. Lark Hill Toyohashi Univ Technol 36:45–60

Yamada M, Owaki Y (2012) Estimation of 4 inter-regional input-output tables for Aichi prefecture for 2005. Chukyo University Institute of Economics, Discussion Paper Series, No. 1205, pp 1–53 (in Japanese)

Yamaguchi M, Shibusawa H (2000) An econometric analysis of East Mikawa area – focusing on the water/supply/demand structure. Stud Reg Sci 31(1):163–179. (in Japanese)

Yamaguchi M, Shibusawa H (2005) A comparative analysis of water resources and regional economic structures in east Mikawa. Stud Reg Sci 35(4):1007–1020. (in Japanese)

Printed by Printforce, the Netherlands